TEACHING NUMBER

Advancing Children's Skills and Strategies

Second Edition

**Robert J. Wright, Jim Martland,
Ann K. Stafford, Garry Stanger**

Paul Chapman Publishing

First published 2002
This edition published 2006

Paul Chapman Publishing Ltd
A SAGE Publications Company
1 Oliver's Yard
55 City Road
London EC1Y 1SP

SAGE Publications Inc
2455 Teller Road
Thousand Oaks
California 91320

SAGE Publications India Pvt Ltd
B42 Panchsheel Enclave
PO Box 4109
New Delhi 110 017

Library of Congress Control Number: 2006922028

A catalogue record for this book is available from the British Library

ISBN 10 1-4129-2184-8   ISBN 13 978-1-4129-2184-8
ISBN 10 1-4129-2185-6   ISBN 13 978-1-4129-2185-5 (pbk)

Typeset by Pantek Arts Ltd, Maidstone, Kent
Printed in Great Britain by Athenaeum Press, Gateshead, Tyne & Wear
Printed on paper from sustainable resources

TEACHING NUMBER

To Sally and Anna

To Mark, Amy, James, Tom and Toby

To Sybil and Jane

To Margaret, Linda and Michael

Contents

List of Figures

List of Photographs

List of Tables

Contributors

AUTHORS

Dr Robert J. Wright holds the position of Professor in Mathematics Education at Southern Cross University in Australia and is an internationally recognized leader in understanding and assessing young children's numerical knowledge and strategies, publishing many articles and papers in this field. His work in the last 15 years has included the development of the Mathematics Recovery Programme which focuses on providing specialist training for teachers to advance the numeracy levels of young children assessed as low attainers. In Australia, the UK, the USA and elsewhere, this programme has been implemented widely and applied extensively to classroom teaching and to average and able learners as well as low attainers. He has conducted several research projects funded by the Australian Research Council including a current project focusing on assessment and intervention for 8- to 10-year-olds.

Jim Martland is a member of the International Board of Mathematics Recovery and Director of Mathematics Recovery Programme (UK) Limited. He is Senior Fellow in the Department of Education at the University of Liverpool. In his long career in education he has held headships in primary and middle schools and was Director of Primary Initial Teacher Training. In all the posts he continued to teach and pursue research in primary mathematics. His current work is with local education authorities in the UK and Canada, delivering professional development courses on assessing children's difficulties in numeracy and designing and evaluating teaching interventions.

Ann Stafford's academic background includes graduate study at Southern Cross University, Australia, the University of Chicago, and Clemson University. She received a Master's degree from Duke University and an undergraduate degree from the University of North Carolina at Greensboro. Her professional experience includes teaching and administrative roles in K-5 classrooms and supervision in the areas of mathematics, gifted, early childhood, and remedial as well as teaching and research positions at Clemson University. She has led in the writing and development of Early Childhood and Mathematics Curricula for the School District of Oconee County, South Carolina. Ann has received numerous professional awards and grants for outstanding contributions to the region and state for mathematics and leadership. She is currently involved in leading the implementation and classroom applications of Mathematics Recovery in the USA.

Garry Stanger has had a wide-ranging involvement in primary, secondary and tertiary education in Australia. He has held positions of Head Teacher, Deputy Principal and Principal and has been a Mathematics Consultant with the New South Wales Department of Education. He has also taught in schools in the USA. He has worked with Robert Wright on the Mathematics Recovery project since its inception in 1992 and has been involved in the development of the Count Me In Too early numeracy project. He currently teaches Master's courses in early numeracy at Southern Cross University.

CONTRIBUTOR TO CHAPTER 1

Peter Gould is the Chief Education Officer in mathematics with the New South Wales Department of Education and Training. He has been instrumental in designing the Count Me In and Count Me In Too projects. His major interest is the effective use of research in the design and delivery of mathematics education.

Acknowledgments

This book is a culmination of several interrelated projects conducted over the past 15 years, many of which come under the collective label of the Mathematics Recovery and Count Me In Too projects in Australia, the UK and the USA. All these projects have involved one or more of the authors undertaking research, development and implementation in collaboration with teachers, schools and school systems. These projects have received significant support from the participating schools and school systems.

The authors wish to express their sincere gratitude and appreciation to all the teachers, students and project colleagues who have participated in and contributed to these projects. We also wish to thank the following organizations for funding and supporting one or more projects which have provided a basis for writing this book: the government and Catholic school systems of the north coast region of New South Wales, Australia, the Australian Research Council, and the New South Wales Department of Education and Training; the School District of Oconee County and the South Carolina State Department of Education, and many other school districts across the USA; the University of Liverpool and the Wigan, Sefton, Salford, Stockport, Knowsley and Cumbrian educational authorities in England; Flintshire County Council in Wales; the University of Strathclyde, and the Glasgow, West Dunbartonshire, Edinburgh and Stirling education authorities in Scotland; Frontier School Division in the Province of Manitoba, Canada; and the Ministry of Education in the Bahamas.

Series preface

If you ask educationalists and teachers whether numeracy intervention deserves equal attention with literacy intervention the overwhelming answer is 'Yes, it should'. If you then ask whether this happens in their experience the answer is a resounding 'No!' What then are the reasons for this discrepancy? Research shows that teachers tend to regard addressing difficulties in literacy as more important than difficulties in early numeracy. Teachers also state that there is a lack of suitable tools for assessing young children's numeracy skills and knowledge and appropriate programmes available to address the deficits.

The three books in this series make a significant impact to redress the imbalance by providing practical help to enable schools and teachers to give equal status to early numeracy intervention. The books are:

▶ *Early Numeracy: Assessment for Teaching and Intervention*, 2nd edition, Robert J. Wright, Jim Martland and Ann K. Stafford, January 2006
▶ *Teaching Number: Advancing Children's Skills and Strategies*, 2nd edition, Robert J. Wright, Jim Martland, Ann K. Stafford and Garry Stanger, June 2006
▶ *Teaching Number in the Classroom with 4–8 year-olds*, Robert J. Wright, Garry Stanger, Ann K. Stafford and Jim Martland, March 2006.

The authors are internationally recognized as leaders in the field of early numeracy intervention. They draw on considerable practical experience of delivering training courses and materials on how to assess young children's mathematical knowledge, skills and strategies in addition, subtraction, multiplication and division. This is the focus of *Early Numeracy*. The revised version contains six comprehensive diagnostic assessment tools to identify children's strengths and weaknesses and has a new chapter on how the assessment provides the direction and focus for teaching intervention. *Teaching Number* sets out in detail nine principles which guide the teaching together with 180 practical, exemplar teaching activities to advance children to more sophisticated strategies for solving numeracy problems. The third book, *Teaching Number in the Classroom with 4–8 year-olds*, extends the work of assessment and teaching intervention with individual and small groups to working with whole classes. In this new text the lead authors have been assisted by expert, primary practitioners from Australia, the USA and the UK who have provided the best available instructional activities for each of eight major topics in early number learning.

The three books in this series provide a comprehensive package on

1. The identification, analysis and reporting on children's numerical knowledge, skills and strategies
2. How to design, implement and evaluate a course of teaching intervention
3. How to incorporate both assessment and teaching in the daily numeracy programme in differing class organizations and contexts.

The series is distinctive from others in the field because it draws on a substantial body of recent, theoretical research supported by international, practical application. Because all the assessment and teaching activities portrayed have been empirically tested the books have the additional, important distinction that they indicate to the practitioner ranges of responses and patterns of behaviour which children tend to make.

The book series provides a package for professional growth and development and an invaluable, comprehensive resource for both the experienced teacher concerned with early numeracy intervention and for the primary teacher who has responsibility for teaching numeracy in kindergarten to upper junior level. Primary numeracy consultants, mathematics advisers, special education teachers, teaching assistants and initial teacher trainees around the world will find much to enable them to put numeracy intervention on an equal standing with literacy. At a wider level the series will reveal many areas of interest to educational psychologists, researchers and academics.

Find out more about Math Recovery by visiting our website at http://www.mathrecovery.org

Preface

This book provides an approach to the teaching of early numeracy which is both structured and progressive. It complements and builds on the assessment of children's number knowledge and strategies documented in the authors' *Early Numeracy: Assessment for Teaching and Intervention, 2nd edition* (Sage 2006). This second revised book sets out a distinctive approach to teaching, which has been developed through the Mathematics Recovery and Count Me In Too projects in Australia, the UK, the USA and Canada.

The authors present a comprehensive and integrated framework for assessment, learning and teaching which provides experienced and beginning teachers with a clear direction and purpose based on three sound principles. The first is that teaching must take account of the child's current number knowledge and strategies used to solve problems. Second, we provide a learning framework in number which sets out learning pathways. These provide the teacher with detailed guidance on where to take the child so that more sophisticated strategies are developed. Third, the book presents a structure of principles and procedures for teaching which link to the learning framework.

Following an explanation of the rationale and background to the work there are five chapters (Chapters 5–9), each of which gives the detailed procedures for advancing children's learning across each of five stages of early arithmetical learning – emergent, perceptual, figurative, initial number and facile number. In each of these chapters a typical profile of a child at a particular stage is presented. This is followed by teaching procedures, which are organized into six key topics. Each key topic includes a clearly defined purpose, demonstrated links to the learning framework in number, illustrated teaching procedures closely attuned to the stage and level of the child's ability, examples of children's likely responses, and suggested vocabulary and materials. The key topics are followed by examples of lessons.

Thus the book provides teachers with directionality and purpose based upon clearly identifying where their pupils are, and a knowledge of where to take them.

The principles and procedures appropriate to a constructivist, inquiry-based approach are set out in the chapters which provide guidance on whole-class teaching and on the teaching of individuals and small groups in specialist settings.

This book presents activities and approaches, which have been tried and tested by practitioners on three continents. The framework can be used with each country's numeracy strategy. The book is of interest to all those who are concerned with finding ways to advance children's numerical strategies and to raising standards in schools. Teachers, advisers, numeracy consultants, special needs coordinators and learning support personnel, as well as teacher educators and researchers whose work relates to this field, will find much of interest and of practical help to develop confidence and skill in the teaching of early number.

Introduction

This book has as its focus the important topic of the teaching of number. It describes a teaching approach and related teaching activities that have been developed in the Mathematics Recovery Programme and the Count Me In Too projects in Australia, the UK, the USA and Canada. The central thrust is the teaching of number skills, knowledge and understanding which is present in the number strand of a primary mathematics curriculum.

The authors have a particular concern for those aspects of early number taught in the 4- to 8-year age range. The reason for this is that research studies in the 1990s showed that there are significant differences in the numerical knowledge of children when they begin school (Aubrey, 1993; Wright, 1991a; 1994; Young-Loveridge, 1989; 1991). These studies also show that these differences in number knowledge increase as children progress through schooling. There is a clear tendency for low attainers in the early years to continue to be low attainers throughout their primary years and to develop negative attitudes to mathematics. There is a need therefore to give every child a positive understanding and success in early number work.

But how does one achieve that? Let us look at the teachers' experiences from the three areas involved in the Mathematics Recovery and Count Me In Too projects. Whether in Australia, the UK or the USA we contend that teachers have a similar basic set of questions to address when planning teaching. We would suggest that the following are prominent in teachers' thoughts:

▶ What knowledge does the child possess?
▶ What are the child's current misunderstandings and misconceptions?
▶ Where do I want to take the child?
▶ Given that the child is having difficulty with 'x', which is a necessary part of learning 'y', what are the optimum tasks, examples and settings that will allow the child to progress?
▶ What specific materials will I use?
▶ How can I establish the linkages and generalities in the learning, which will allow progress to be made?
▶ When I provide these experiences what will I look for to be able to gather evidence that the child is learning?
▶ What possible misconceptions might occur and how can I remediate these?

We would contend that the teacher also has other questions to address. A particularly difficult one relates to one's own confidence level in understanding the mathematical content. It is easy to state but hard to confront. It is 'Do I understand the mathematics involved?' Given the nature and content of early numeracy this should not be hard but let us rephrase the question to read: 'First, do I understand in which order I should teach the content and second, how do the aspects of what I am teaching fit together and lead to a greater understanding of the operations in numeracy?' Internationally, many countries and states have focussed energy and resources on the review of the mathematics curriculum, the refinement and articulation of attainment targets for each age group, the production of curriculum materials and resource books for teachers, and the attendant professional development training courses. Consequently, teachers have much clearer guidelines and there has been a growth in teacher confidence and children's attainment.

However, we would contend that not everything is as straightforward as it may seem. The guidance produced suggests targets to be met across year groups. Furthermore suggested, even mandated, styles of teaching have put a greater emphasis on whole-class interaction in the introductory and plenary parts of the numeracy lesson. Even when group work is evident, government reports indicate that children are not being sufficiently challenged. The activities in the teachers' handbooks may be interesting and fun for the children but do they engender learning which relates to the individual child's needs?

Where then, is the help to be given to the teacher of numeracy in the early years who has a class with a wide range of abilities and a significant number of children already beginning to demonstrate low attainment? We think it is a valid question to ask that under pressure to advance a designated proportion of children to a certain level of achievement, what time, help and support can be given to the children who are beginning to fall behind?

WHAT IS SPECIAL ABOUT THIS BOOK?

In writing this book we want to address this problem and provide practical help for the teacher. It is not a book to dip into for interesting exercises, though we provide plenty of these. The authors consider that this book differentiates itself from other books on numeracy because we present a comprehensive and integrated approach to assessment, learning and teaching which is strong on providing teachers with a clear direction and purpose based on three sound principles.

First, we consider that it is crucially important for teaching to take full account of the child's current numerical knowledge and strategies used to solve problems. The term 'strategies' is used here to refer to procedures that children might use to solve an addition task such as 8 + 4 or a division task such as 'Twelve biscuits are divided between some children. Each child gets two biscuits each. How many children received the biscuits?' The term 'numerical knowledge' includes aspects such as identifying written symbols for numbers (for example, 7, 35) and knowing sequences of number words such as 'one, two, three … '.

Second, we present a learning framework for number, that is, a framework that links closely to the assessment tasks and sets out children's learning pathways for teachers. An example of the framework in action can be seen when a young child is asked to solve 5 + 4. The child says the forward number word sequence 'one, two, three, four, five' and raises the fingers on one hand sequentially in coordination with the number words. The child continues 'one, two, three, four' and raises four fingers on the other hand sequentially. The child then counts the raised fingers from one, 'one, two, three, four, five, six, seven, eight, nine, … Nine!' The above could well be accompanied by touching the face with each raised finger in turn.

In the example above the child used a relatively low level strategy which we describe as 'counting from one'. The Learning Framework in Number, which is explained in full in Chapter 1, provides detailed guidance on how the teacher can advance the child so that the use of finger patterns becomes more sophisticated or redundant and the child now routinely uses a more advanced strategy.

Thirdly, we present a set of principles and procedures for teaching that are closely linked with the learning framework. We demonstrate how a child can progress from one level of strategy to the next more advanced level, by illustrating the learning framework in action using six key topics, each of which spawns a multitude of practical and motivating teaching activities.

WHAT IS THE BACKGROUND TO THIS BOOK?

The approach presented is drawn from a programme of numeracy research and development projects which have been implemented widely in schools in several countries since the early 1990s. Many of these implementations have taken place under the label of Mathematics Recovery. The Mathematics Recovery Programme involves an extensive study of school-based and team-based professional development. It aims to advance teachers' knowledge of the assessment, learning and teaching of early number.

The initial development of Mathematics Recovery occurred over a three-year period (1992–95) in the Australian state of New South Wales. The development was funded by the Australian Research Council and participating school systems. Since 1995, the Mathematics Recovery Programme has been widely implemented in 24 states in the USA, and in the Bahamas and Canada. In total this has involved approximately 350 specialist teachers, 60 leaders of training and more than 3,000 participating children. In England and Wales 21 education authorities have implemented Mathematics Recovery on a sustainable basis and in Scotland and Ireland over 200 teachers at all levels have been trained in assessment and intervention.

Furthermore, in a recent comprehensive study on the help available to children with numeracy difficulties for the Department of Education and Skills for England and Wales (Dowker, 2004), the Mathematics Recovery Programme was identified as one of only two large-scale independently developed, individualized intervention programmes which specifically address numeracy difficulties in young children. The research report indicated how Mathematics Recovery targeted the younger primary age group, 4–8-year-olds, drew together research findings regarding the stages of development and, most importantly, made them applicable to teaching and teachers of early numeracy. The report highlighted how children in the programme showed significant improvement even after short interventions of between 10 and 12 hours' duration and how the children reached, or even exceeded, the age-related norms for their classmates. The report also indicated that teachers saw their work in the Mathematics Recovery Programme as being a very important part of their professional development, enabling them to be more confident in diagnosing children's difficulties in early numeracy, to design and implement effective interventions and to advise colleagues on courses of action.

In 1996 the Mathematics Recovery Programme was adapted by the state department of education in New South Wales as the basis of a systemic initiative in mathematics in the early years of schooling and was called Count Me In Too (CMIT). In its initial year, CMIT was piloted in 13 schools. In subsequent years (1997 to 2004) CMIT was progressively implemented in virtually all of the 1,700 schools (primary/elementary) across the state. In addition, CMIT has been widely adopted by school systems in other Australian states, and in 2000 CMIT was the basis of a nationally funded pilot project in New Zealand involving 81 schools (Thomas and Ward, 2001). This pilot of CMIT, together with the Mathematics Recovery Programme, informed the New Zealand Numeracy Development Project which was implemented nationally from 2001 onward (Bobis et al., 2005).

A list of selected publications relating to both the Mathematics Recovery Programme and Count Me In Too is included in the Bibliography.

PURPOSE OF THIS BOOK

In the revised edition of *Early Numeracy: Assessment for Teaching and Intervention* (Sage 2006), the authors set out a detailed approach to the assessment of children's early numerical strategies and knowledge. The book provides six diagnostic interview schedules for focussing on a range of aspects of early number. The schedules are administered via videotaped assessment interviews and used to elicit children's strategies and number knowledge. As well, the book sets out procedures for analyzing the results of the assessment interviews, and a framework for determining a child's strategies and documenting current levels of a child's knowledge. The framework just referred to is called the Learning Framework in Number (LFIN).

The approach to assessment just described has been used by a range of school systems in at least five countries, as part of the Mathematics Recovery project or programmes based on the Count Me In Too initiative. This work has involved several thousand teachers and many thousands of children. The current authors and their colleagues have spent countless hours conducting school-based and system-based professional development meetings which have focussed specifically on this approach to assessment, and more generally on ways of observing and interpreting children's mathematical activity and thinking. In these meetings we are frequently asked by teachers and leaders to provide advice about teaching approaches that accord with children's mathematical thinking. Many of the questions in this vein relate to a specific episode or event which the participants have all just observed via a videotaped replay of assessment or teaching, because videotaped records of assessment and teaching sessions constitute an important means by which to observe practice during the professional development meetings.

One might respond to questions of the kind just described in general terms by referring the audience to the LFIN, in which case the teachers who ask the question (that is, about teaching) would be expected to refer to the framework and attempt to determine appropriate instructional approaches. For many of the teachers with whom we have worked this kind of approach has worked well. In Mathematics Recovery for example, teachers are provided with an extensive bank of instructional settings and activities, as well as videotapes exemplifying the use of these. Another way to respond when asked what teaching is appropriate given a particular episode or event in an assessment or teaching session is to select or describe a setting and demonstrate teaching procedures and learning activities. Again, this approach has been used frequently by the current authors in Mathematics Recovery professional development meetings.

The purpose of this book is to answer systematically the questions described above, that is, questions relating to the need for teaching to be informed by assessment and to accord with the LFIN. The book provides detailed descriptions of specific instructional settings, teaching procedures and learning activities. These descriptions are organized in terms of the LFIN and relate closely to the outcomes of the assessment procedures set out in the related books *Early Numeracy: Assessment for Teaching and Intervention* (Sage 2000) and *Early Numeracy: Assessment for Teaching and Intervention, Second Edition,* (Sage 2006).

THE STRUCTURE OF THE BOOK

In this introduction we have described the scope of this book, that is, the teaching of early number and we have outlined what is special about this book. We have also introduced the reader briefly to two interrelated initiatives which have had a significant impact internationally in the area of early number learning. These are the Mathematics Recovery Programme and the New South Wales Count Me In Too early numeracy initiative. Further, we have provided an international overview of recent issues and developments in mathematics in the early years of school. Finally, we have referred to our previous book which sets out a comprehensive approach to assessment in early number and describes a comprehensive learning framework which guides assessment and provides a means of observing and documenting the outcomes of assessment.

Chapter 1 consists of an introductory section and two main sections. In the introductory section we describe how we use terms such as 'strategies' and 'knowledge' in describing children's early number learning. In the first main section we describe the Learning Framework in Number (LFIN) which provides essential guidance for teaching as well as for assessing. This section includes links from the LFIN to key topics and teaching procedures in Chapters 5–9. The second main section describes aspects of our approach to teaching early number.

From Chapter 2 onwards the reader is provided with a comprehensive guide to teaching early number using the LFIN. This guide includes a focus on individualized teaching as it might be used, for example, in an intervention programme, as well as a focus on classroom teaching. The purpose of Chapter 2 is to provide for the reader, a detailed overview of significant aspects of the approach to individualized teaching which has been developed in the Mathematics Recovery Programme. These aspects are organized into the following three sections: Section A – the guiding principles of individualized teaching; Section B – key elements of individualized teaching; and Section C – characteristics of children's problem-solving in individualized teaching sessions. The final section of Chapter 2 consists of four scenarios of children's learning in individualized teaching sessions. These scenarios serve to exemplify the aspects described in Sections A, B and C.

In Chapter 3 key attributes of whole-class teaching of number are discussed. A variety of lesson formats is described. The chapter first describes the fundamental instructional goals of sense-making and intellectual autonomy. The chapter then focusses on the teaching and learning cycle which is described in four sections: 'Where are they now?', 'Where do you want them to be?', 'How will they get there?' and 'How will you know when they've arrived?'

Chapter 4 provides a detailed overview of the common format that is used in Chapters 5–9. This includes an overview of the 30 key instructional topics that are presented in Chapters 5–9, that is, six in each chapter. Each key topic consists of around six teaching procedures. In this way, each of the five chapters includes at least 32 teaching procedures. The description of each teaching procedure includes examples of a teacher's words, actions, and notes on purpose, teaching and children's responses.

Chapters 5–9 set out a detailed framework for teaching which relates closely to and is informed by the LFIN. Each of these five chapters focusses on teaching children at a given stage in terms of the Stages of Early Arithmetical Learning. In this vein Chapter 5 focusses on teaching children at the Emergent Stage, Chapter 6 focusses on the Perceptual Stage, Chapter 7on the Figurative Stage, Chapter 8 on the Stage of Counting-On and Down-From, and the Stage of Counting-Down-To, and Chapter 9 on the Facile Stage. As well, each of these stages assumes specific levels of knowledge in relation to other aspects of the LFIN, namely, forward number word sequences (FNWSs), backward number word sequences (BNWSs), numeral identification and tens and ones.

Chapters 5–9 have a common structure. First, an overview is provided of the early number knowledge and strategies typical of children at the stage and levels specified for that chapter. This is followed by detailed outlines of six key teaching topics considered likely to be significant for advancing the knowledge of such children. The key topics relate to a range of important aspects of the LFIN, and each outline of a key topic includes descriptions of, on average, six teaching procedures relevant to that key topic. Instructional materials, key vocabulary and links to the LFIN are included for each key topic, and each teaching procedure includes notes on purpose, teaching and children's responses. Each of Chapters 5–9 also includes outlines of several lessons which exemplify whole-class teaching to children assumed to be at the stage and levels specified for that chapter.

Across Chapters 5–9 there is a total of 182 teaching procedures. This very large corpus of teaching procedures, each of which can be understood in terms of the LFIN, is a particularly distinctive feature of this book. The authors' intention is that the reader should regard the teaching procedures as illustrative and should adopt and adapt procedures as they see fit. The procedures are not intended to be followed verbatim. To do so would not accord with current practice in Mathematics Recovery teaching sessions, nor would it accord with the guiding principles of individualized teaching set out in Chapter 2 of this book.

1

Advancing Children's Strategies and Knowledge in Early Number

This chapter consists of an introductory section and two main sections. In the introductory section we describe how we use terms such as 'strategies' and 'knowledge' in describing children's early number learning. In the first main section we describe the Learning Framework in Number which guides our assessment and the ways we observe and describe children's learning in early number. Finally, the second main section focusses on our approach to teaching early number.

CHILDREN'S STRATEGIES AND KNOWLEDGE

This section provides an explanation of how we use terms such as 'strategies' and 'knowledge' in describing how we understand and characterize children's learning of early number.

Children's Strategies in Early Number

In referring broadly to children's learning of early number we find it useful to use the terms 'strategies' and 'knowledge'. The term 'strategies' refers to the procedures that a child might use to solve various kinds of early number tasks. Of course, there is a multitude of strategies used by children in early number learning, and any particular child might use a wide range of strategies. One can also determine differences among strategies in terms of their mathematical sophistication. Related to this, advancement in the sense of developing more sophisticated strategies is an important part of learning mathematics. Thus it follows that, in studying early number learning, it is important to focus on the strategies children use. At the same time we believe that a focus on children's strategies alone cannot lead to fully understanding early number learning. Progression in early number learning should not be viewed only as involving the development of new strategies that an adult might classify as 'more efficient' than earlier strategies. Our view is that knowing the strategies that a child uses to solve early number tasks provides an important indication of the current levels of the child's learning but does not comprehensively describe the child's learning.

Children's Early Number Knowledge

We use the term 'knowledge' to encompass aspects of children's early number learning that we consider somewhat separate from the children's strategies. The child, who can easily read 2-digit numerals for example, has developed important early number knowledge. This is a straightforward example of an aspect of the child's knowledge that cannot simply be described or captured by describing a strategy the child uses to solve a problem.

Using the Term 'Knowledge' in an All-Encompassing Sense

In our work focussing on learning and teaching early number we use the term 'knowledge' in two senses. The first of these is the one just described, that is, aspects of children's learning that are not easily characterized as strategies. Second, we use the term 'knowledge' in a broad and all-encompassing sense. Put simply, the knowledge that this child has in the area of early number consists of everything that the child knows about early number. This seemingly circular definition of 'knowledge' (that is, defining the child's knowledge as what the child knows) is useful we believe. Thus the numerical knowledge of a 7-year-old might include aspects such as: (a) using advanced counting strategies (counting-on and counting-back) to solve additive and subtractive tasks; (b) reading and writing all 2-digit numerals and many 3-digit numerals; (c) knowing some basic addition facts (for example, doubles); and (d) counting by twos, fives and tens.

Knowledge and Learning

'Knowledge' used in the sense just described differs from that seen in some curriculum documents where 'knowledge' is used as one of three terms to encompass learning, that is 'knowledge, skills and understandings'. Thus our use of 'knowledge' is likely to encompass skills and understandings. The use of the term 'understand' as in, for example, 'the child understands addition' is problematic we believe, simply because there are many levels of understanding of mathematical processes such as addition. Finally, we do not focus on a distinction between 'concepts' and 'procedures'. Our view is that, at least in the area of early number learning, concepts and procedures are closely interrelated. We view a child's concept of addition for example, primarily in terms of the strategies or procedures that the child uses in additive situations. Thus from our point of view, it makes sense to talk about a child's knowledge of addition. This might include the range of strategies the child might use in additive situations and the sense the child might make of the symbol '+'. What follows, which is critical for teaching, is to learn as much as possible about the child's current knowledge in early number and to do this it is necessary to observe closely children's words and actions in appropriate mathematical contexts. Observing children for the purpose of learning about their current number knowledge often requires the teacher to decentre, that is, to step out of their current framework for viewing mathematics and its learning. Finally, our view of knowledge and learning in early number is strongly constructivist (von Glasersfeld, 1995), and we advocate a problem-based or inquiry-based approach to teaching (Cobb and Bauersfeld, 1995).

THE LEARNING FRAMEWORK IN NUMBER

In the Introduction we described how the ideas presented in this book are based on our work with literally thousands of teachers on three continents. In working with these teachers we first introduce them to an approach to assessment and the LFIN which provides a blueprint for the assessment and indicates likely paths for children's learning. In this section we provide a brief description of key aspects of the LFIN. These are described in more detail in the authors' previous book, that is, the revised edition of *Early Numeracy: Assessment for Teaching and Intervention* (*ENATI*) (Wright et al., 2006). As well, many of these ideas are revisited from a perspective of instruction, in later chapters of this book. An overview of the LFIN is provided in Table 1.1.

Table 1.1 The Learning Framework in Number

Part A	Part B	Part C	Part D
Early Arithmetical Strategies Base-Ten Arithmetical Strategies	Forward Number Word Sequences and Number Word After Backward Number Word Sequences and Number Word Before Numerals	Other Aspects of Early Arithmetical Learning	Early Multiplication and Division
Stages: **Early Arithmetical Strategies** 0 Emergent Counting 1 Perceptual Counting 2 Figurative Counting 3 Initial Number Sequence 4 Intermediate Number Sequence 5 Facile Number Sequence	**Levels: Forward Number Word Sequences (FNWS) and Number Word After** 0 Emergent FNWS 1 Initial FNWS up to 'ten' 2 Intermediate FNWS up to 'ten' 3 Facile with FNWSs up to 'ten' 4 Facile with FNWSs up to 'thirty' 5 Facile with FNWSs up to 'one hundred'	Combining and Partitioning Spatial Patterns and Subitizing Temporal Sequences Finger Patterns	**Levels:** 1 Initial Grouping 2 Perceptual Counting in Multiples 3 Figurative Composite Grouping 4 Repeated Abstract Composite Grouping 5 Multiplication and Division as Operations
Levels **Base-Ten Arithmetical Strategies** 1 Initial Concept of Ten 2 Intermediate Concept of Ten 3 Facile Concept of Ten	**Levels: Backward Number Word Sequences (FNWS) and Number Word Before** 0 Emergent BNWS 1 Initial BNWS up to 'ten' 2 Intermediate BNWS up to 'ten' 3 Facile with BNWSs up to 'ten' 4 Facile with BNWSs up to 'thirty' 5 Facile with BNWSs up to 'one hundred' **Levels: Numeral Identification** 0 Emergent Numeral Identification 1 Numerals to '10' 2 Numerals to '20' 3 Numerals to '100' 4 Numerals to '1000'	Base-Five (Quinary-Based) Strategies	

LFIN – an Overview

The LFIN contains 11 aspects of early number learning organized into four parts. Parts A, B and D contain six aspects, each of which is presented in tabular form. These tabulated forms – referred to as models – set out stages or levels of children's numerical knowledge and facilitate profiling of children's knowledge. The 11 aspects of the LFIN should not be regarded as early number topics which are widely separated from each other. Rather, they are closely interrelated. Thus assessing one aspect typically provides information about other aspects. Also, teaching typically focusses on several aspects simultaneously rather than just one aspect.

LFIN: Part A

Aspect A1: Stages of Early Arithmetical Learning (SEAL)

The SEAL sets out a progression of the strategies children use in early number situations which are problematic for them, for example being required to figure out how many in a collection, and various kinds of additive and subtractive situations. The SEAL model appears in Table 1.2 and consists of a progression of five stages in children's development of early arithmetical strategies. The label 'Stage 0' is used for children who have not attained the first stage. The SEAL model has been adapted from research by Steffe and colleagues (for example, Steffe, 1992; Steffe and Cobb, 1988; Steffe et al., 1983) and related research by Wright (1989; 1991a). The SEAL model is considered the primary or most important aspect of the LFIN. Extensive descriptions and examples of the SEAL are available in the previous book (*ENATI, 2nd edition*) and so are not repeated here. Teaching procedures relating to this aspect of the LFIN appear in Key Topics 5.3 (Chapter 5), 6.3 (Chapter 6), and 7.3 (Chapter 7).

Table 1.2 Model for Stages of Early Arithmetical Learning (SEAL)

Stage 0: Emergent Counting. Cannot count visible items. The child either does not know the number words or cannot coordinate the number words with items.

Stage 1: Perceptual Counting. Can count perceived items but not those in screened (that is concealed) collections. This may involve seeing, hearing or feeling items.

Stage 2: Figurative Counting. Can count the items in a screened collection but counting typically includes what adults might regard as redundant activity. For example, when presented with two screened collections, told how many in each collection, and asked how many counters in all, the child will count from 'one' instead of counting-on.

Stage 3: Initial Number Sequence. Child uses counting-on rather than counting from 'one', to solve addition or missing addend tasks (e.g. 6 + x = 9). The child may use a count-down-from strategy to solve removed items tasks (e.g. 17 – 3 as 16, 15, 14 – answer 14) but not count-down-to strategies to solve missing subtrahend tasks (e.g. 17 – 14 as 16, 15, 14 – answer 3).

Stage 4: Intermediate Number Sequence. The child counts-down-to to solve missing subtrahend tasks (e.g. 17 – 14 as 16, 15, 14 – answer 3). The child can choose the more efficient of count-down-from and count-down-to strategies.

Stage 5: Facile Number Sequence. The child uses a range of what are referred to as non-count-by-ones strategies. These strategies involve procedures other than counting-by-ones but may also involve some counting-by-ones. Thus in additive and subtractive situations, the child uses strategies such as compensation, using a known result, adding to ten, commutativity, subtraction as the inverse of addition, awareness of the 'ten' in a teen number.

Aspect A2: Base-Ten Arithmetical Strategies

Around the time they attain Stage 3, 4 or 5 on the SEAL, children typically begin to develop knowledge of the tens and ones structure of the numeration system. Of course, children can and should solve addition and subtraction tasks involving 2-digit numbers (that is from 10 onward) long before they develop knowledge of the tens and ones structure. For children who have attained Stage 5, development of knowledge of the tens and ones structure becomes increasingly important. Table 1.3 outlines a progression of three levels in children's development of base-ten arithmetical strategies. The model for the development of base-ten arithmetical strategies is adapted from research by Cobb and Wheatley (1988). Teaching procedures relating to this aspect of the LFIN appear in Key Topics 8.3 (Chapter 8), 9.3 (Chapter 9) and 9.4 (Chapter 9).

LFIN – Part B

The three aspects in Part B are concerned with important specific aspects of children's early number knowledge: forward number word sequences, backward number word sequences, and numeral identification. These models resulted from research by Wright (for example, 1991b; 1994).

Table 1.3 Model for development of base-ten arithmetical strategies

---

Level 1: Initial Concept of Ten. The child does not see ten as a unit of any kind. The child focusses on the individual items that make up the ten. In addition or subtraction tasks involving tens, children count forward or backward by ones.

Level 2: Intermediate Concept of Ten. Ten is seen as a unit composed of ten ones. The child is dependent on re-presentations (like a mental replay or recollection) of units of ten such as hidden ten-strips or open hands of ten fingers. The child can perform addition and subtraction tasks involving tens where these are presented with materials such as covered strips of tens and ones. The child cannot solve addition and subtraction tasks involving tens and ones when presented as written number sentences.

Level 3: Facile Concept of Ten. The child can solve addition and subtraction tasks involving tens and ones without using materials or re-presentations of materials. The child can solve written number sentences involving tens and ones by adding or subtracting units of tens and ones.

Note: A necessary condition for attaining Level 1 is attainment of at least Stage 3 in the Stages of Early Arithmetical Learning.

Aspects B1 and B2 – FNWSs and BNWSs

The term 'number words' refers to the spoken and heard names of numbers. In the LFIN an important distinction is made between counting and reciting a sequence of number words. This distinction was made by Steffe et al. (1983). The term 'counting' is used only in cases which involve coordination of each spoken number word with an actual or imagined (that is, conceptualized) item. Thus counting typically occurs in situations which are problematic for children, for example solving an additive or subtractive problem or establishing the numerosity of a collection of items. The activity of merely saying a sequence of number words is not referred to as counting.

The term 'forward number word sequence' refers to a regular sequence of number words forward, typically but not necessarily by ones, for example the FNWS from one to twenty, the FNWS from eighty-one to ninety-three, the FNWS by tens from twenty-four. The term 'backward number word sequence' is used in similar vein, for example the BNWS from twenty to ten. From the point of view of fully understanding children's early numerical knowledge, it is useful to construe as distinct, the two aspects concerned with number word sequences, that is forward and backward. Nevertheless, because of the many similarities between these two aspects their presentations here are integrated to some extent. Models associated with these aspects are shown in Tables 1.4 and 1.5. The label 'Level 0' is used for children who have not attained the first level. Finally, teaching procedures relating to these aspects of the LFIN appear in Key Topics 5.1 (Chapter 5), 6.1 (Chapter 6) and 7.1 (Chapter 7).

Table 1.4 Model for the construction of forward number word sequences

---

Level 0: Emergent FNWS. The child cannot produce the FNWS from 'one' to 'ten'.

Level 1: Initial FNWS up to 'ten'. The child can produce the FNWS from 'one' to 'ten'. The child cannot produce the number word just after a given number word in the range 'one' to 'ten'. Dropping back to 'one' does not appear at this level. Children at Levels 1, 2 and 3 may be able to produce FNWSs beyond 'ten'.

Level 2: Intermediate FNWS up to 'ten'. The child can produce the FNWS from 'one' to 'ten'. The child can produce the number word just after a given number word but drops back to 'one' when doing so.

Level 3: Facile with FNWS up to 'ten'. The child can produce the FNWS from 'one' to 'ten'. The child can produce the number word just after a given number word in the range 'one' to 'ten' without dropping back. The child has difficulty producing the number word just after a given number word, for numbers beyond ten.

Level 4: Facile with FNWS up to 'thirty'. The child can produce the FNWS from 'one' to 'thirty'. The child can produce the number word just after a given number word in the range one' to 'thirty' without dropping back. Children at this level may be able to produce FNWSs beyond 'thirty'.

Level 5: Facile with FNWSs up to 'one hundred'. The child can produce FNWSs in the range 'one' to 'one hundred'. The child can produce the number word just after a given number word in the range 'one' to 'one hundred' without dropping back. Children at this level may be able to produce FNWSs beyond 'one hundred'.

Table 1.5 Model for the construction of backward number word sequences

---

Level 0: Emergent BNWS. The child cannot produce the BNWS from 'ten' to 'one'.

Level 1: Initial BNWS up to 'ten'. The child can produce the BNWS from 'ten' to 'one'. The child cannot produce the number word just before a given number word. Dropping back to 'one' does not appear at this level. Children at Levels 1, 2 and 3 may be able to produce BNWSs beyond 'ten'.

Level 2: Intermediate BNWS up to 'ten'. The child can produce the BNWS from 'ten' to 'one'. The child can produce the number word just before a given number word but drops back to 'one' when doing so.

Level 3: Facile with BNWSs up to 'ten'. The child can produce the BNWS from 'ten' to 'one'. The child can produce the number word just before a given number word in the range 'one' to 'ten' without dropping back. The child has difficulty producing the number word just before a given number word, for numbers beyond ten.

Level 4: Facile with BNWSs up to 'thirty'. The child can produce the BNWS from 'thirty' to 'one'. The child can produce the number word just before a given number word in the range 'one' to 'thirty' without dropping back. Children at this level may be able to produce BNWSs beyond 'thirty'.

Level 5: Facile with BNWSs up to 'one hundred'. The child can produce BNWSs in the range 'one hundred' to 'one'. The child can produce the number word just before a given number word in the range 'one' to 'one hundred' without dropping back. Children at this level may be able to produce BNWSs beyond 'one hundred'.

Aspect B3 – Numeral Identification

Numerals are the written and read symbols for numbers, for example '3', '27', '360'. Learning to identify, recognize and write numerals can rightly be regarded an important part of early literacy development. At the same time it is important to realize that this learning is equally, if not more so, an important part of early numerical development. The term 'identify' is used here with precise meaning, that is, to state the name of a displayed numeral. The complementary task of selecting a named numeral from a randomly arranged group of displayed numerals is referred to as 'recognizing'. Thus we make the distinction between 'numeral identification' and 'numeral recognition'. Using these terms in this way accords with typical use in psychology and in early literacy. Table 1.6 outlines a progression of four levels in children's development of numeral identification. As with the models above, the label 'Level 0' is used for children who have not attained the first level. Finally, teaching procedures relating to this aspect of the LFIN appear in Key Topics 5.2 (Chapter 5), 6.2 (Chapter 6), 7.2 (Chapter 7) and 8.2 (Chapter 8).

Table 1.6 Model for the development of numeral identification

---

Level 0: Emergent Numeral Identification. Cannot identify some or all numerals in the range '1' to '10'.
Level 1: Numerals to '10'. Can identify numerals in the range '1' to '10'.
Level 2: Numerals to '20'. Can identify numerals in the range '1' to '20'.
Level 3: Numerals to '100'. Can identify one and two digit numerals.
Level 4: Numerals to '1000'. Can identify one, two and three digit numerals.

LFIN – Part C

For the purposes of the LFIN, the aspects in Part C are not presented in tabular form, that is, the LFIN does not contain models for any of these aspects. Nevertheless, these aspects are considered important in children's early number learning and they often arise incidentally in concert with the aspects in Parts A and B. Thus these aspects are closely interrelated with those in Parts A and B. There are also close interrelationships among the aspects within Part C. Descriptions of each of these parts are now provided.

Aspect C1: Combining and Partitioning

Combining and partitioning refers to those aspects of children's early number knowledge that involve combining or partitioning small numbers (typically in the range 1–10), and do not involve counting-by-ones. Consider the following scenario: a child is asked 'what are three and three?' and the child immediately answers 'six'. We might conclude that this child can combine three and three without counting, and thus, in some sense, the child knows the combination 'three and three make six'. The sense in which the child knows the combination might involve other aspects in Part C, such as spatial patterns or finger patterns. Consider a second scenario: a child is asked to say two numbers that make 7 when added and the child immediately answers 'five and two'. In similar vein to the above, we might conclude that this child can partition seven into five and two without counting-by-ones. We include Combining and Partitioning in the LFIN because development of this aspect of early number knowledge can form an important basis for further learning (see Cobb et al., 1991; 1992; McClain and Cobb, 1999; Yackel et al., 1991). Teaching procedures relating to this aspect of the LFIN appear in Key Topics 7.4 and 7.5 (Chapter 7).

Aspect C2: Spatial Patterns and Subitizing

Spatial patterns and subitizing refers to those aspects of children's early number knowledge that involve spatial configurations of various kinds, for example domino patterns, pairs patterns, ten frame, playing card patterns, regular plane figures and random arrays. Subitizing is 'the immediate, correct assignation of number words to small collections of perceptual items' (von Glasersfeld, 1982). Activities involving spatial patterns and subitizing have an important role in advancing young children's number knowledge, and there is a range of instructional settings in which subitizing can feature prominently. Teaching procedures relating to this aspect of the LFIN appear in Key Topics 5.4 (Chapter 5) and 6.4 (Chapter 6).

Aspect C3: Temporal Sequences

Temporal sequences refers to those aspects of children's early number knowledge associated with events that occur sequentially in time, for example sequences of sounds and movements. Children's early number knowledge can be advanced via activities based on temporal sequences. Examples of such activities are counting or copying rhythmical sequences of sounds, and counting monotonic sequences of sounds or movements. Teaching procedures relating to this aspect of the LFIN appear in Key Topics 5.6 (Chapter 5).

Aspect C4: Finger Patterns

Children's use of finger patterns in early number contexts is extremely widespread. Fingers are used in a range of ways and with varying levels of sophistication. One child might solve an additive task such as 5 + 3 by sequentially raising five fingers on one hand while counting from one to five, sequentially raising three fingers on the other hand while counting from one to three, and then counting all the raised fingers from one to eight. Another child might solve the same task by raising three fingers sequentially while counting from six to eight. The view taken by the authors of this book is that instruction in early number must accord with and take account of children's spontaneous finger-based strategies. Through appropriate instruction children should develop more sophisticated finger-based strategies, for example, strategies that are conceptually based rather than perceptually based. Ultimately, the child should progress to levels where finger-based strategies are not their main means of proceeding in situations involving addition, subtraction, multiplication or division. Teaching procedures relating to this aspect of the LFIN appear in Key Topics 5.5 (Chapter 5) and 6.5 (Chapter 6).

Aspect C5: Base-Five (Quinary) Strategies

Base-five (quinary) strategies refers to those aspects of children's early number knowledge that involve using the number five as a base. These arise in instructional settings involving the arrangement of items in fives, for example the arithmetic rack and the ten frame, and also settings where use of fingers is prominent (because there are five fingers on each hand). Using five as a base means that combining and partitioning of numbers involving five is given special emphasis in children's learning, and is therefore likely to be incorporated by children into additive and subtractive strategies. Typically in these settings, the number ten is also a base, for example two rows of five can be seen as ten. Thus five does not replace ten as a base. Rather, five is an additional base. In early number there is a major potential advantage associated with five being used as an additional base along with ten. This is because using five as a base has the potential to greatly reduce reliance on counting-by-ones. Development of this aspect has drawn on research by Gravemeijer (for example, 1994) and Cobb and colleagues (Cobb et al., 1995; Gravemeijer et al., 2000). Finally, teaching procedures relating to this aspect of the LFIN appear in Key Topics 7.4 and 7.5 (Chapter 7) and 8.5 (Chapter 8).

LFIN – Part D: Early Multiplication and Division

The LFIN just described was developed and used in the early years of Mathematics Recovery and Count Me In Too. In more recent years the LFIN has been applied more broadly to early number learning, and was extended to include a focus on young children's development of multiplication and division knowledge. The extension of LFIN to multiplication and division drew on a range of research (see Mulligan, 1998). For a detailed description of assessment tasks for early multiplication and division the reader is referred to the authors' previous book, that is, the revised edition of *ENATI* (Wright et al., 2006). The results of this extension to multiplication and division are now described. In similar vein to the aspects in Parts A and B of the LFIN, the multiplication and division aspect is presented in tabular form consisting of five levels. As before, these are regarded as levels of progression and are used to profile children's knowledge (Mulligan, 1998). The model appears in Table 1.7 and, in the following section, each level is explained in more detail. Finally, teaching procedures relating to this aspect of the LFIN appear in Key Topics 6.6 (Chapter 6), 7.6 (Chapter 7), 8.6 (Chapter 8) and 9.6 (Chapter 9).

Level 1: Initial Grouping

The child at Level 1 can establish the numerosity of a collection of equal groups when the items are visible and counts by ones when doing so, that is the child uses perceptual counting (see Stage 1 of SEAL). The child can make groups of a specified size from a collection of items, for example given 12 counters the child can arrange the counters into groups of three thereby obtaining four groups. This is referred to as quotitive sharing, and is also known as the grouping aspect of division. Also, the child can share a collection of items into a specified number of groups, for example given 20 counters the child can share the counters into five equal groups. This is referred to as partitive sharing and is also known as the sharing aspect of division. The child does not count in multiples.

Table 1.7 Model for early multiplication and division levels

Level 1: Initial Grouping. Uses perceptual counting (that is, by ones) to establish the numerosity of a collection of equal groups, to share items into groups of a given size (quotitive sharing) and to share items into a given number of groups (partitive sharing).

Level 2: Perceptual Counting in Multiples. Uses a multiplicative counting strategy to count visible items arranged in equal groups.

Level 3: Figurative Composite Grouping. Uses a multiplicative counting strategy to count items arranged in equal groups in cases where the individual items are not visible.

Level 4: Repeated Abstract Composite Grouping. Counts composite units in repeated addition or subtraction, that is, uses the composite unit a specified number of times.

Level 5: Multiplication and Division as Operations. Can regard both the number in each group and the number of groups as a composite unit. Can immediately recall or quickly derive many of the basic facts for multiplication and division.

Level 2: Perceptual Counting in Multiples

The child at Level 2 has developed counting strategies that are more advanced than those used in Level 1. These multiplicative counting strategies involve implicitly or explicitly counting in multiples. After sharing a collection into equal groups the child uses one of these strategies to count all the items contained in the groups which are necessarily visible. The child is not able to count the items in situations where the groups are screened. These counting strategies include rhythmic, double and skip counting, and each is given the label 'perceptual' (for example, perceptual rhythmic counting) because of the child's reliance on visible items.

Level 3: Figurative Composite Grouping

The child at Level 3 has developed counting strategies which do not rely on items being visible and which do not involve counting-by-ones. For example, if the child is presented with four groups of three counters, where each group is separately screened, the child may use skip counting by threes to determine the number of counters in all, that is, 'three, six, nine, twelve'. From the child's perspective each of the four screens symbolizes a collection of three items but the individual items are not visible. There is a correspondence between not having to count by ones on tasks involving equal groups and counting-on in the case of an additive task, for example 6 + 3 presented with two screened collections. In the case of the additive task the first screen symbolizes the collection of six counters and the child does not need to count from one to six. Thus one would expect children at Level 3 in terms of early multiplication and division to have attained Stage 3 in terms of SEAL.

Level 4: Repeated Abstract Composite Grouping

The child at Level 4 has constructed a conceptual structure labeled an 'abstract composite unit' (Steffe and Cobb, 1988) in which the child is simultaneously aware of both the composite and unitary aspects of three for example. The child can use repeated addition to solve multiplication tasks and repeated subtraction to solve division tasks, and can do so in the absence of visible or screened items. On a multiplicative task involving six groups of three items, in which each group is separately screened, the child is aware of each group as an abstract composite unit. Construction of abstract composite units is associated with Stage 5 on SEAL. Thus, in the case of children who have significant experience with multiplication and division situations as well as addition and subtraction situations, it is likely that being at Level 4 on early multiplication and division is contemporaneous with being at Stage 5 in terms of SEAL.

Level 5: Multiplication and Division as Operations

The child at Level 5 can coordinate two composite units in the context of multiplication or division. On a task such as six threes or six groups of three for example, the child is aware of both six and three as abstract composite units, whereas at Level 4, the child is aware of three as an abstract composite unit but is not aware of six as an abstract composite unit. The child at Level 5 can immediately recall or quickly derive many of the basic facts of multiplication and division, and may use multiplication facts to derive division facts. At Level 5, the commutative principle of multiplication (for example, $5 \times 3 = 3 \times 5$) and the inverse relationship between multiplication and division are within the child's zone of proximal development (ZPD). Thus for example, the child might be aware that six threes is the same as three sixes and might use $4 \times 8 = 32$ to work out $32 \div 4$.

TEACHING EARLY NUMBER

This section consists of four parts. In the first part we provide an overview of the teaching of 2- and 3-digit addition, subtraction and place value. This part is included because it addresses important issues concerning the teaching of early number which do not arise in the earlier section of this chapter which focusses on the Learning Framework in Number. Thus this part focusses on the teaching of the most advanced aspects of early number and the teaching of children at the most advanced stage and levels of the LFIN. As well, this overview focusses on recent developments in the teaching of early number. These recent developments are incorporated into key teaching topics presented in Chapters 5–9. In the second part in this section we present the Instructional Framework in Early Number (IFEN). The IFEN is organized into three strands: (a) counting, (b) grouping and (c) number words and numerals. The IFEN consists of a progression of key teaching topics which are elaborated in detail in Chapters 5–9. These key topics take account of the results of assessment and are informed by a consideration of the LFIN. In the third part of this section we provide two classroom scenarios to exemplify the teaching of early number. These scenarios also exemplify topics from the first and second parts of this section. Finally, in the fourth part we explain how the IFEN links closely to the LFIN.

Teaching of 2- and 3-Digit Addition and Subtraction and Place Value

In the 1970s and 1980s there was an accepted and dominant approach to teaching 2- and 3-digit addition and subtraction and the related topic of place value associated with 2- and 3-digit numbers, which we shall call here, the 'traditional approach'. Use of the traditional approach has extended well into the 1990s but its use is now being increasingly challenged. As well, the traditional approach was the dominant one in several countries, for example, in the USA, Australia and the UK where the approach is predominant in many of the popular text series of the 1980s and 1990s. In the traditional approach a central idea is that children are first taught place value concepts, and this is undertaken using base-ten materials of some kind. Typically, children complete a lot of exercises that involve writing the numeral that corresponds to an arrangement using base-ten materials or vice versa, for example the child writes '35' to correspond to three longs (that is, tens) and five ones. Teaching place value in this way is followed by teaching standard algorithms, again using base-ten materials. Thus children might use base-ten materials to symbolize 38 and 46, and follow a procedure with the materials that corresponds more or less with the standard columnar written procedure for adding numbers, that is, the standard vertical algorithm for written addition. In similar vein, base-ten materials would be used in correspondence with the standard written algorithm for subtraction and so on.

Challenges to the Traditional Approach

Since the mid-1980s at least, the traditional approach has been challenged from several viewpoints (for example, Cobb, 1991; Treffers, 1991, pp. 27–30). Many of these take issue with either the specific way in which base-ten materials are used in the traditional approach or the idea of teaching place value prior to adding or subtracting 2-digit numbers and so on. Kamii (1986) observed that '(w)hen children are prematurely given place value instruction, they sometimes build two systems separately, and juxtapose a system of ones and an independent system of tens' (ibid., p. 78). Kamii advocated an approach that contrasts with the traditional approach in which children invent their own procedures for adding 2- and 3-digit numbers, prior to formal teaching of place value. Another challenge to the traditional approach has come with the increasing emphasis on teaching mental strategies and developing children's mental arithmetic (Beishuizen and Anghileri, 1998). Related to this, is the view that teaching the standard algorithms too early (that is, in first, second or third grade) interferes with the further development of mental arithmetic, particularly in the case of lower-attaining children (Beishuizen and Anghileri, 1998, p. 522; Treffers, 1991, p. 48).

Emphasizing the Development of Children's Mental Strategies

Beishuizen and Anghileri (1998, pp. 520–1) highlight the distinction between mental recall and mental strategies, which together constitute what is usually called mental arithmetic. Mental strategies refers to informal mental methods of calculating, while mental recall is linked to habituation or memorization in the context of single-digit arithmetic, that is, what is frequently referred to as learning the basic facts. Children who are strong in mental recall, that is, they have a well-developed set of memorized facts, are better equipped to develop increasingly advanced mental strategies (Gray, 1991, pp. 569–70). Thus mental recall is an important basis for the development of mental strategies. As Beishuizen and Anghileri advise, the development of mental strategies should not be left to chance – teaching should be intentionally directed to the development of increasingly advanced mental strategies. It is not sufficient simply to foster and promote the use of any strategies.

Categorizing Mental Strategies for 2-Digit Addition and Subtraction – Jumping or Splitting

Beishuizen (1993, p. 395) used the labels N10 and 1010 to signify two categories of strategies. N10, also known as 'the jump method' involves the child adding or subtracting from a number, for example, solving 56 + 31 by first adding 30 (or 10 and 10 and 10) to 56, and finally adding 1. 1010, also known as 'the split method' involves working separately with the tens and ones, for example, solving 56 + 31 by first adding 50 and 30, and then adding 6 and 1 and so on. Several other research studies have similarly described these two categories of children's strategies or cognitive orientations. Thus the N10 and the 1010 categories correspond respectively with (a) children's counting-based and collections-based interpretations of 2-digit numerals (Cobb and Wheatley, 1988); (b) the cumulative sums and partial sums categories of additive strategies (Thompson, 1994, p. 333); (c) sequence-tens and separate-tens types of strategies (Fuson et al., 1997); and (d) the incrementing and combining tens and ones types of invented algorithms for multi-digit addition and subtraction (Carpenter et al., 1999, pp. 70–3).

NIO or 1010 – Child Preference

Thompson (1994) found that, in a particular context, children had a preference for 1010 strategies. However, Beishuizen and Anghileri (1998) argue from research studies that N10 strategies should generally be emphasized. In those studies children initially, and in particular lower-attaining children, were observed to have a preference for 1010 strategies. This occurs because children initially find it difficult to increment by tens off the decade which is inherent in N10 (for example, to count 37, 47, 57, and so on). Nevertheless, when they encounter the more difficult problems – 2-digit addition and subtraction with carrying (that is, regrouping/borrowing) – the children using 1010 are much more likely to make errors and find it more difficult to appropriately modify their strategies. Therefore it is argued that N10 strategies should be promoted initially, and this can be adjusted to an approach integrating N10 and 1010 as the children progress in terms of the sophistication of their strategies.

The Empty Number Line

The empty number line (ENL) is an instructional context or setting which arose out of curriculum research and development projects in the Netherlands, which are labeled Realistic Mathematics

Education (RME) (for example, Gravemeijer, 1994; Streefland, 1991). In the ENL approach, a simple line is drawn and the children's task is to use appropriate notation to demonstrate their mental strategy. For the task 45 + 32, for example, the ENL is empty except for an initial mark with the accompanying numeral (that is, 45). The child might draw one large arc to indicate an increment of 30 or three small arcs indicating increments of 10, and so on. Eventually children become skilled at using the ENL in this way to demonstrate their particular strategy for solving a task. The ENL constitutes a context for discussion of and reflection on strategies. Over the course of several lessons or several weeks it is expected that children's methods become more sophisticated. They might begin to increment or decrement by several tens rather than only one ten, might develop strategies involving compensation, and might use 1010 kinds of strategies in conjunction with N10 strategies.

Place Value – How and When

Beishuizen and Anghileri (1998) argue for a different approach to the teaching of place value from that described above (that is, the traditional approach) and take issue with UK writers who specify that 'place value still should remain a central principle for [these] mental methods' (ibid., p. 531). They argue that '[t]he ENL introduction of early number is not based on the teaching of the *place value concept* in the first place but develops more gradually through the extension of *mental strategies*' (emphasis in original) (ibid., p. 522). This view of teaching place value gradually, and in the context of developing mental strategies, corresponds exactly with the approach advocated by several writers in the USA, notably Kamii (1986) and Cobb and Wheatley (1988). As well, this view is quite at odds with the traditional approach as outlined above.

Counting-Based and Collection-Based Approaches

Classroom teaching experiments conducted and reported by Cobb and colleagues (for example, Cobb and Bauersfeld, 1995; Cobb et al., 1991; 1992; Yackel et al., 1991) in the 1990s provide useful models for teaching 2- and 3-digit addition and subtraction and place value. These methods are closely aligned with the Realistic Mathematics Education approach (see above). The counting-based approach relates closely to the ENL context. Children are introduced to the ENL and its use for recording strategies that might involve initially, incrementing or decrementing 2-digit numbers by tens and ones. Around the same time as they are working with the ENL, children can be presented with problems in the context of tens and ones materials (for example, bundling sticks or multi-base arithmetic blocks (MAB), that is, base-ten blocks). Tasks presented to children in this context can involve enumerating – that is, saying the number of blocks in 3 bundles of ten and six ones. This can be followed by tasks involving incrementing and decrementing by tens and ones. In this approach the technique of screening can be used judiciously to encourage mental strategies (rather than perceptual counting and counting-by-ones), in much the same way that screening of collections of counters is used when initially teaching counting, adding and subtracting involving smaller numbers.

A Balanced Approach to Multi-Digit Arithmetic

In this book we advocate a balanced approach to 2- and 3-digit addition and subtraction and place value. This approach draws extensively on the work of Cobb and colleagues (for example, Cobb et al., 1997a; 1997b), and in turn draws on aspects of the Realistic Mathematics Education approach in the Netherlands (for example Gravemeijer, 1994; Streefland, 1991). As well, we believe the approach takes account of many of the issues raised by Beishuizen and Anghuleri (1998). The approach taken here involves a balance of two approaches. Cobb and Wheatley (1988) labeled these as 'counting-based' and 'collections-based'. This approach places strong emphasis on the gradual development of mental strategies for 2-digit addition and subtraction, and through this the gradual development of place value knowledge. This approach is set out in several of the Key Topics in Chapters 8 and 9 (see 8.3, 9.2 and 9.3).

Instructional Framework for Early Number

In this section we provide an overview of the IFEN, which links closely to the LFIN described previously in this chapter. The IFEN consists of a progression of key teaching topics and is organized into three strands: (a), number words and numerals, (b) counting and (c) grouping (see below). It is organized into five phases of early number instruction. The IFEN appears in Table 1.8.

Table 1.8 Instructional Framework for Early Number

Instructional phase	Number words and numerals	Counting	Grouping
Phase 1 **Emergent**	• FNWS 1 to 20 • BNWS 1 to 1 0 • Numerals 1 to 10	• Counting involving visible items in collections and in rows • Temporal sequences and temporal patterns	• Early spatial patterns • Early finger patterns
Phase 2 **Perceptual**	• FNWS 1 to 30 • BNWS 1 to 30 • Numerals 1 to 20	• Counting involving screened items in collections and rows	• Developing spatial patterns • Developing finger patterns • Equal groups and sharing
Phase 3 **Figurative**	• FNWS 1 to 100 • BNWS 1 to 100 • Numerals 1 to 100	• Counting-on and counting-back to solve additive and subtractive tasks	• Combing and partitioning using 5 and 10 • Combing and partitioning in range 1–10 • Early multiplication and division
Phase 4 **Counting-on and Counting-back**	• NWS by 2s, 10s, 5s, 3s, and 4s, in the range 1–100 • Numerals 1 to 1000	• Incrementing by 10s and 1s	• Adding and subtracting to and from decade numbers • Adding and subtracting to 20 using 5 and 10 • Developing multiplication and division
Phase 5 **Facile**	• NWS by 10s on and off the decade • NWS by 100s on and off the 100 and on and off the decade	• 2-digit addition and subtraction through counting	• 2-digit addition and subtraction involving collections • Non-canonical forms of 2- and 3-digit numbers • Higher decade addition and subtraction • Advanced multiplication and division

Strands of the IFEN

The three strands of IFEN are intended to be broad and inclusive of specific early number topics. Thus the IFEN is intended to be inclusive of topics such as addition, subtraction, multiplication, division and place value knowledge. The IFEN is inclusive of topics such as knowing combinations to 10, knowing the basic facts (that is, number bonds), learning to derive facts from known facts, and mental strategies for 2-digit addition and subtraction. Our intention is that these three strands should come close to encompassing a typical early number curriculum. These strands are not necessarily distinct from each other. Rather, they are considered to be overlapping to some extent, that is, some specific aspects of early number could relate to key topics in more than one of the strands. In this section each of these strands is described and illustrations of particular aspects of children's knowledge are provided.

Number words and numerals   The term 'number words' is used in the IFEN to refer in the main, to the spoken and heard names for numbers (for example, 'six', 'twenty-seven'). As well, the terms 'forward number word sequence' and 'backward number word sequence' are used to refer to number words when spoken or heard in the standard sequence (for example, one, two, three … and ten, nine, eight, …). In the view of the authors, developing facility with number words and numerals is also a very important aspect of early number learning. Aspects to be developed include saying the FNWS and the number word after a given number; saying the BNWS and the number word before; and saying number word sequences by twos, tens, fives, threes and fours, and by tens on and off the decade. This area also includes numeral identification, numeral recognition, activities based on numeral sequences, activities involving the sequencing of numerals (involving numerals in a standard sequence – 11, 12, … 20), and ordering numerals (involving numerals not in a standard sequence – 8, 11, 17, 21).

Examples of children's knowledge relating to number words and numerals are:

▶   When asked to say the number after six, a child says, 'one, two, three, four, five, six, seven!'
▶   When asked to count from 27 to 33, a child says, 'twenty-seven, twenty-eight, twenty-nine, twenty-ten . . .'.
▶   When asked to write the numeral 'sixteen', a child writes a '6' and then writes a '1' to the left of the '6'.
▶   When asked to name the numeral '207', a child says 'twenty and seven'.

Counting   The term 'counting' is used in the IFEN to refer to aspects of young children's numerical thinking that involve counting by ones (rather than, for example, by threes). 'Counting' is used in situations where we assume the child has a cognitive goal, for example to determine the numerosity of a collection. Thus the child is attempting to solve a problem and in doing so is conceptualizing items of some kind, rather than merely reciting the number word sequence. Thus counting involves a different cognitive activity from that involved when saying the number word sequence. A more extensive discussion of counting has already been provided previously in this chapter (see Aspect A1: Stages of Early Arithmetical Learning).

Examples of children's strategies that illustrate counting are:

▶   When asked to get 14 counters from a pile, a child counts out the counters one by one in coordination with saying the words from 'one' to 'fourteen'.
▶   When asked to solve an additive task such as 9 + 3 presented with two screened collections, a child counts on from nine to 12 in coordination with raising three fingers.
▶   When asked how many counters are removed from a collection, given that there were 24 and now there are only 19, a child solves the problem by counting-down-to and keeping track of the number of counts (that is, twenty- three, twenty-two, … nineteen, five!).

Grouping   The term 'grouping' is used in the IFEN to refer to aspects of young children's numerical thinking that involve numbers larger than one, and that involve procedures other than counting-by-ones. Grouping in this sense is relevant to the development of strategies for adding, subtracting, multiplying and dividing, and for the development of knowledge of tens and ones and place value. Given this description we can choose to see grouping as juxtaposed with and complementary to counting (in the sense of counting-by-ones).

Examples of children's strategies that illustrate grouping are:

▶ When asked 'what are two and two', a child raises two fingers simultaneously on one hand, and then two simultaneously on the other and says 'four', assuming that the child did not count by ones from one or count-on from two by ones.

▶ When asked 15 – 11 a child says '15 take away 10 is 5 and you just take one away, 4'.

▶ When asked to work out three fives a child says '5, 10, 15!'

Phases of the IFEN

The IFEN is organized into five phases of early number instruction. These relate to SEAL in the LFIN. Phase 1 concerns the child who is at the Emergent Stage (Stage 0) and progressing to the Perceptual Stage (Stage 1). Phase 2 concerns the child who is at the Perceptual Stage (Stage 1) and progressing to the Figurative Stage (Stage 2). Phase 3 concerns the child who is at the Figurative Stage (Stage 2) and progressing to Stages 3 and 4 on SEAL. Phase 4 concerns the child who is at Stage 3 or 4 and progressing to Stage 5. Finally, Phase 5 concerns the child who is at Stage 5.

Exemplifying the Teaching of Early Number

This section presents two classroom scenarios which are included to provide the reader with initial examples of the teaching of early number. (Further discussion and illustrations of the teaching of early number are provided from Chapter 2 onwards.) These scenarios exemplify three topics discussed earlier in this section:

1. Counting and grouping strategies – Classroom Scenario 1
2. N10 (that is, the jump method) – Classroom Scenario 2
3. The instructional setting of the empty number line (ENL) – Classroom Scenarios 1 and 2.

Classroom Scenario 1. 1

In this excerpt from a first-grade class the teacher uses the ENL to symbolize the strategies children use to solve an additive task involving a collection of eight counters and a collection of four counters. The strategies include counting from one (Ben), counting-on (Meg) and the grouping strategy of adding through ten (Amy).

Teacher: (Displays and then screens eight counters.) Here are eight counters. I'll put this cover over them. (Places four counters next to the screen.) Here are another four counters. How many counters are there altogether? (Points to Ben.)

Ben: I said one, two, three, four, five, six, seven, eight. Then I went nine, ten, eleven, twelve. There are twelve.

Teacher: (Draws on the ENL to symbolize Ben's strategy as shown in Figure 1.1.) Meg, tell us how you did it.

Meg: I think the answer is eleven. I said eight, nine, ten, eleven.

Teacher: (Draws an ENL.) How many did we start with Meg?

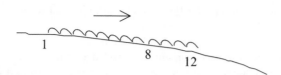

Figure 1.1 Classroom Scenario 1.1

Figure 1.2 Classroom Scenario 1.1

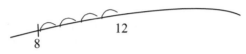

Figure 1.3 Classroom Scenario 1.1

Meg: Umm – eight.

Teacher: (Writes a mark on the ENL and writes '8' below the mark. See Figure 1.2.) Now how many steps do we need to take?

Meg: Four.

Teacher: Meg would you please come to the board and show us the four steps.

Meg: (Comes to the board. Starting at eight, she marks the four steps as shown in Figure 1.3.)

Teacher: So Meg you started at eight, then made four steps. Say the numbers out aloud and we'll follow.

Meg: I started at eight. Then (pointing to each hop in turn) there's nine, ten, eleven, twelve. The answer's twelve.

Teacher: Would someone else please tell us how he or she worked it out?

Amy: I just knew it was twelve. It's easy. You see I can make ten and I know that ten and two is twelve. So the answer's twelve.

Teacher: Did you see what Amy did? She used two of the counters to make a group of ten, then she knows that ten and another two makes twelve. What would ten and another six be? Ten and four? Ten and seven? (and so on.)

Classroom Scenario 1.2

In this excerpt from a second grade class the teacher uses the ENL to symbolize the strategies children use to solve the task of 54 take away 18.

Teacher: What is 54 take away 18? (Allows time for most children to solve the problem.) Darcy, you seemed to get an answer very quickly. Please tell us what you did. I'll try to draw your solution on the board. (Draws an ENL on the board. See Figure 1.4.)

Darcy: I came back 10, that's 44. Then I came back four, which is 40, then another 4 which makes 36.

Teacher: Great. Are there any questions?

Stefan: I did the same. I came back 10 to get 44 but then I counted back 8 to get 36. I kept track on my fingers.

Teacher: (Draws an ENL on the board with notation as shown in Figure 1.5.) Thanks Stefan. Are there any comments about these two ways of working out the question?

Paul: I can see they started off the same but Darcy's way is a bit faster. There aren't as many steps. Look (points to the teacher's ENLs on the board). You can see it on the maps on the board.

Teacher: Does anyone have another way of working it out? (Mary raises her hand.) Yes Mary.

Mary: I've got a really fast way. Remember the other day we were coming back ten then twenty then thirty, well I did it like that.

Teacher: What do you mean Mary?

Mary: Well, I came back 20 because 18 is close to 20. Then I added on 2 to get the answer.

Teacher: (Draws an ENL on the board with notation as shown in Figure 1.6.) Is this what you mean Mary? (Points to the ENL on the board.)

Mary: Yes. That's what I did and I think it's quicker than Darcy's way! (Another child raises her hand.)

Teacher: Yes Summer.

Summer: I started at 18. I need two more to make 20. Then from 20 I need 34 to get to 54. So I need 36 to get from 18 to 54. The answer is 36.

Teacher: (Draws an ENL on the board with notation as shown in Figure 1.7.)

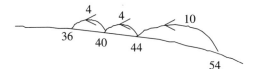

Figure 1.4 Classroom Scenario 1.2

Figure 1.5 Classroom Scenario 1.2

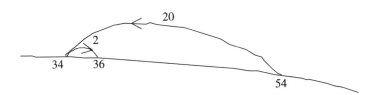

Figure 1.6 Classroom Scenario 1.2

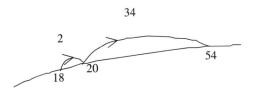

Figure 1.7 Classroom Scenario 1.2

Strategies in Classroom Scenario 1.2   Darcy's strategy involves first partitioning 18 into 10 and 8, then decrementing 54 by 10, then partitioning 8 into 4 and 4, then decrementing 44 by 4 and then 40 by 4. Stefan's strategy initially was similar to Darcy's. But he counted back 8 from 44 by ones. Mary's strategy involves decrementing 54 by 20 and then incrementing 34 by 2. Finally, Summer begins at 18. She increments 18 by 2, and 20 by 34 and keeps track of the numbers by which she increments (that is, 2 and 34).

Linking the Instructional Framework (IFEN) to the Learning Framework (LFIN)

As explained above, the IFEN consists of a progression of key teaching topics. For the sake of brevity these are referred to as Key Topics. Chapters 5–9, contain extensive elaborations of each of these Key Topics. The elaborations include descriptions of teaching procedures relevant to the Key Topic. Typically there are from four to eight teaching procedures for each Key Topic.

Many of the Key Topics in Table 1.8 can be linked to particular stages and levels on one of the models of LFIN (see earlier in this chapter), in the sense that they are considered appropriate exemplars or starting points for advancing children at the lower level or stage on the relevant model. Thus, for example, the Key Topic of 'FNWS to 30' which appears in Phase 2 in the Number Words and Numerals strand (Table 1.8) links directly to Levels 3 and 4 on the model of FNWS (see Table 1.4). This is because the teaching procedures in the Key Topic of FNWS to 30 are considered appropriate for advancing children from Level 3 to Level 4 on the model of FNWS (Table 1.4). In similar vein, the Key Topic of 'Counting involving visible items in collections and in rows' which appears in Phase 1 of the Counting strand (Table 1.8) links directly to Stages 0 and 1 on the model of SEAL (Table 1.2). This is because the teaching procedures in the Key Topic are considered appropriate for advancing children from Stage 0 to Stage 1 of SEAL. As a third example, the Key Topic of 'Incrementing by 10s and 1s' (Phase 4 of the Counting strand) links to Level 2 and Level 3 on the model of development of base-ten arithmetical strategies (Table 1.3). Finally, the Key Topic of 'equal groups and sharing' (see Phase 2 of the Grouping strand) and those relating to multiplication and division (see Phases 3–5 of the Grouping strand) link directly to the model for early multiplication and division (see Table 1.7).

Some of the Key Topics in Table 1.8 do not link directly to a stage or level on one of the models of the LFIN. Rather, the Key Topic links directly to an aspect in Part C of LFIN (see Table 1.1). Thus, for example, the Key Topic of 'Combining and partitioning in the range 1–10' (Phase 3 of the Grouping strand) links directly to Aspect C1 of the LFIN (see earlier in this chapter). In similar vein, the Key Topics of 'Early spatial patterns' (Phase 1 in the Grouping strand) and 'Developing spatial patterns' (Phase 2 in the Grouping strand) link directly to Aspect C2 of the LFIN.

Some of the Key Topics in Phases 4 and 5 necessarily do not link to stages or levels on models in the LFIN. This is because these Key Topics focus on the teaching of the most advanced topics of early number. These Key Topics have the purpose of advancing children who already are near or at the highest levels on the models of LFIN. Many of these Key Topics relate closely to topics discussed above under the heading 'Teaching of 2- and 3-digit addition and subtraction and place value'. Thus the Key Topic of '2-digit addition and subtraction involving collections' (Phase 5 of the Grouping strand) relates to 1010 or 'the split method' and Collections-based approaches described in that section. In similar vein, the Key Topic of '2-digit addition and subtraction through counting' relates to N10 or 'the jump method' and the empty number line.

As a final point, some of the Key Topics do not correspond directly to LFIN for reasons other than those discussed above. Typically these Key Topics focus on specific aspects of early number content which link indirectly with the LFIN. Examples of these Key Topics are 'adding and subtracting to and from decade numbers' (Phase 4 of the Grouping strand) and 'non-canonical forms of 2- and 3-digit numbers' (Phase 5 of the Grouping strand).

2

Individualized Teaching in Mathematics Recovery

Mathematics Recovery (MR) is a programme in which teachers provide intensive, individualized instruction to low-attaining first-grade children (typically 6- or 7-year-olds). This programme of instruction occurs in daily sessions of 30 minutes' duration, for periods of 12 to 14 weeks. The Mathematics Recovery Programme was developed over a three-year period during 1992–95, and since the beginning of its development the programme has involved hundreds of teachers on three continents. In all the implementations of MR, teachers have routinely videotaped their teaching sessions, and the videotaped records of teaching sessions have been used in professional development meetings, and in one-to-one meetings between teacher and leader. Thus videotapes of teaching sessions and of excerpts of teaching constitute a rich source of data for describing and understanding individualized teaching in early number, as it occurs in MR teaching sessions.

The purpose of this chapter is to describe and illustrate key features of MR instruction. Descriptions are provided of 30 features in all, and these are organized into three categories as follows: Section A focuses on the nine guiding principles of MR teaching; Section B focusses on 12 key elements of MR teaching; and Section C focusses on nine characteristics of children's problem-solving in MR teaching sessions.

These key features are followed by descriptions and discussions of illustrative excerpts from MR teaching sessions. The excerpts are referred to as scenarios and are based on videotaped records of MR teaching sessions. For each feature described in Sections A, B or C, reference is made to one or more of the scenarios when that feature is particularly in evidence in the scenario. Table 2.1 sets out for each feature in Sections A, B or C, the scenarios where that feature is evident.

SECTION A – THE GUIDING PRINCIPLES OF MR TEACHING

The individualized teaching sessions afford an opportunity to provide intensive, high-quality teaching. In order to do this the teacher must have a clear model of the child's current knowledge and strategies in early number, and a clear idea of the progress in learning that is a reasonable goal for the child. The origins of MR teaching lie in large part in research projects which involved longitudinal observation and study of children's developing strategies and learning, as they occurred in interactive teaching sessions. This teaching approach can be characterized by the following set of nine guiding principles.

Table 2.1 Scenarios in which guiding principles of teaching, key elements of teaching and characteristics of children's problem-solving are evident

No.	Description/topic	Scenario			
		1	2	3	4
Section A – Guiding principles of MR teaching					
1	Problem-based/inquiry-based teaching	x	x	x	x
2	Initial and ongoing assessment	x	x	x	x
3	Teaching just beyond the cutting edge (ZPD)	x	x	x	x
4	Selecting from a bank of teaching procedures	x	x	x	x
5	Engender more sophisticated strategies	x	x	x	x
6	Observing the child and fine-tuning teaching	x	x	x	x
7	Incorporating symbolizing and notating				x
8	Sustained thinking and reflection	x	x	x	x
9	Child intrinsic satisfaction	x	x	x	x
Section B – Key elements of MR teaching					
1	Micro-adjusting	x	x	x	x
2	Scaffolding	x		x	x
3	Handling an impasse	x			
4	Introducing a setting		x		
5	Pre-formulating a task		x	x	
6	Reformulating a task			x	
7	Post-task wait-time	x	x	x	x
8	Within-task setting change	x			
9	Screening, color-coding and flashing	x	xxx	xx	
10	Teacher reflection	x	x	x	x
11	Child checking	x	x	x	x
12	Affirmation	x	x	x	x
Section C – Characteristics of children's problem-solving					
1	Cognitive reorganization			x	x
2	Anticipation	x		x	
3	Curtailment				x
4	Re-presentation			xxx	
5	Spontaneity, robustness and certitude	x	x	x	x
6	Asserting autonomy				x
7	Child engagement	x	x	x	x
8	Child reflection	x	x	x	x
9	Enjoying the challenge	x			x

1. *The teaching approach is inquiry based, that is, problem based. Children routinely are engaged in thinking hard to solve numerical problems which, for them, are quite challenging.*
2. *Teaching is informed by an initial, comprehensive assessment and ongoing assessment through teaching. The latter refers to the teacher's informed understanding of the child's current knowledge and problem-solving strategies, and continual revision of this understanding.*
3. *Teaching is focussed just beyond the 'cutting-edge' of the child's current knowledge.*

One-to-one teaching

4. *Teachers exercise their professional judgment in selecting from a bank of teaching procedures each of which involves particular instructional settings and tasks, and varying this selection on the basis of ongoing observations.*
5. *The teacher understands children's numerical strategies and deliberately engenders the development of more sophisticated strategies.*
6. *Teaching involves intensive, ongoing observation by the teacher and continual micro-adjusting or fine-tuning of teaching on the basis of his or her observation.*
7. *Teaching supports and builds on the child's intuitive, verbally based strategies and these are used as a basis for the development of written forms of arithmetic which accord with the child's verbally based strategies.*
8. *The teacher provides the child with sufficient time to solve a given problem. Consequently, the child is frequently engaged in episodes which involve sustained thinking, reflection on her or his thinking and reflecting on the results of her or his thinking.*
9. *Children gain intrinsic satisfaction from their problem-solving, their realization that they are making progress and from the verification methods they develop.*

Each of these principles is now discussed in more detail and, where appropriate, links are made from the principle to one or more of the scenarios in Section D.

Principle 1

The teaching approach is inquiry based, that is problem based. Children routinely are engaged in thinking hard to solve numerical problems which for them, are quite challenging.

Individualized teaching involves presenting tasks and challenges to children which are problematic for them. It is through solving challenging tasks that children's knowledge is advanced. In these situations children can reorganize their current ways of construing and solving tasks. Of course, teaching in this way is not limited to the MR Programme. Enquiry-based or problem-based teaching is typical in teaching approaches with a constructivist orientation, that is, the learner is actively engaged in processes of constructing knowledge. This principle is exemplified in each of the four scenarios.

Principle 2

Teaching is informed by an initial, comprehensive assessment and ongoing assessment through teaching. The latter refers to the teacher's informed understanding of the child's current knowledge and problem-solving strategies, and continual revision of this understanding.

Assessment plays a critical role in individualized teaching. The initial assessment provides a comprehensive profile of the child's current knowledge and strategies. The initial assessment enables determination of the current levels of the child's knowledge in terms of SEAL, levels of FNWS development, and so on. But, importantly, the initial assessment provides rich and precise detail about the particular strategies currently used by the child and their current knowledge relating to number words and numerals. The initial assessment is continually updated through the course of the teaching sessions. The teaching sessions routinely involve problem-solving on the part of the child. And during this problem-solving an important responsibility for the teacher is to review and update their hypothesized model of the extent of the child's current knowledge and strategies. In each of the four scenarios, the teacher is closely observing the child. This process of close observation is the basis from which the teacher continually updates his or her assessment of the child.

Principle 3

Teaching is focussed just beyond the 'cutting-edge' of the child's current knowledge.

Vygotsky, a Russian educational researcher who wrote in the 1930s, is attributed with the idea of a learner's 'zone of proximal development' (Blanck, 1990, p. 50; Vygotsky, 1978). This idea has much currency today in various areas of education. For Vygotsky, the zone of proximal development relates to knowledge that the learner is capable of learning under the influence of appropriate teaching, and this zone is regarded as more extensive than that consisting of the knowledge that the learner is capable of learning without assistance. In individualized teaching the teacher has a clear purpose of extending the child's current knowledge. Thus, focussing just beyond the cutting-edge is akin to teaching in the child's zone of proximal development (ZPD). It is necessary for the teacher in this situation to be clearly aware of what the child currently knows and what new learning is reasonably within the child's grasp through intensive teaching. In each scenario, it is evident that the problems posed by the teacher are genuine problems for the child. At the same time the child has a good likelihood of solving problems, and on many occasions, the child's success can be attributed, at least in part, to aspects of the interactive teaching. This illustrates the notion of zone of proximal development (ZPD).

Principle 4

Teachers exercise their professional judgment in selecting from a bank of teaching procedures each of which involves particular instructional settings and tasks, and varying this selection on the basis of ongoing observations.

Each teaching session in MR involves using several (for example, four to six) teaching procedures. Teaching procedures typically consist of (a) a setting, that is, a device (for example, a numeral track) or materials (for example, bundling sticks); (b) tasks that are presented to the child; and (c) a series of steps undertaken to present the tasks. Chapters 5–9 contain many examples of teaching procedures.

These teaching procedures are intended to be indicative and illustrative rather than prescriptive. Thus the procedures typify instructional approaches that are likely to be productive at particular points in children's learning. It is important for the teacher to learn to judiciously select, use and modify the teaching procedures. The teaching procedures in the scenarios are adapted from those in Key Topics (Chapters 5–9) as follows: Scenarios 1 and 4 from Key Topic 8.3, Scenario 2 from Key Topic 7.4, and Scenario 3 from Key Topic 7.3.

Principle 5

The teacher understands children's numerical strategies and deliberately engenders the development of more sophisticated strategies.

This principle relates to strategies such as counting, adding and subtracting, used by children to solve numerical tasks. An important part of teachers' learning is to become very familiar with these strategies, and to understand them in the context of the LFIN. Teachers should maintain a strong interest in extending their understanding of children's means of problem-solving. Teachers should also regard each instance of genuine problem-solving on the child's part as new and distinct, and as an opportunity to extend their current knowledge of the nature and range of children's strategies and interrelated number knowledge. Teachers who develop their own rich, experientially based model of the likely ways that children at particular points in the LFIN solve problems, are well equipped to undertake the task of engendering more sophisticated problem-solving behavior in their children. This principle is evident in each of the four scenarios. In Scenario 1, for example, Kathryn is endeavoring to engender strategies of incrementing and decrementing by tens and ones without using materials, and in Scenario 3 Jane is trying to engender the strategy of counting-on to solve missing addend tasks.

Principle 6

Teaching involves intensive, ongoing observation by the teacher and continual micro-adjusting or fine-tuning of teaching on the basis of his or her observation.

In the discussion of Principle 4 the notion of a 'teaching procedure' was introduced. Teaching procedures indicate productive ways of instructing children but they do not provide the fine detail of interactive teaching. That fine detail relates to the teacher's moment by moment monitoring of the child's responses and changing of the task or issue of current focus. This moment by moment monitoring and changing is referred to as 'micro-adjusting', and this is a key characteristic of successful individualized teaching. Fine-tuning of teaching is evident in each scenario. In Scenario 3, for each new missing addend task, Jane carefully chooses the numbers in the task. In Scenario 4, when Wendy places out the eighth bundle she deliberately refrains from saying how many bundles in all.

Principle 7

Teaching supports and builds on the child's intuitive, verbally based strategies and these are used as a basis for the development of written forms of arithmetic which accord with the child's verbally based strategies.

For children in the 5- to 7-year-old range, much of their initial number learning involves verbally based strategies. By this we mean that their cognitive activity when engaged in solving number tasks typically involves sound images of number words. Consider, for example, a child who counts from one to solve an additive task such as 6 + 3 presented with two screened collections. We hypothesize that, for most children at least, their mental activity is largely concerned with spoken and heard (that is, as distinct from written number words) number words and thus sound images of numbers words are important in the child's awareness. This holds true, we believe, in cases where the task might involve reading or writing numerals. An important aspect of MR teaching is to incorporate ways of symbolizing and notating that support children's verbally based strategies. Symbolizing and notating in this way, provides a means not only to record strategies, but to extend the current range of strategies (Gravemeijer et al., 2000). Extension of strategies can involve an incorporation of symbolizing into current strategies. Thus, increasingly, the child develops strategies in which means of symbolizing are within the children's awareness when using the strategy. This principle is exemplified in Scenario 4 where Joanna's task includes using plastic digits to form decade numbers corresponding to the number of bundling sticks.

Principle 8

The teacher provides the child with sufficient time to solve a given problem. Consequently, the child is frequently engaged in episodes which involve; sustained thinking, reflection on her or his thinking and reflecting on the results of her or his thinking.

Providing instruction in individualized teaching sessions is quite different from the kinds of teaching that usually occur in regular classroom teaching. For many teachers, providing an intensive 30-minute period of instruction involves significant change and adjustment to familiar ways of operating. One important change in teaching approach relates to giving children sufficient time to think about challenging problems. Classroom teaching typically involves interacting with several children simultaneously or in quick succession. In such circumstances the teacher is usually very aware of competing needs to communicate in subtly differing ways with different children. Individualized teaching does not involve aspects such as children's competing needs for attention. Thus working effectively in an individualized teaching session requires adjustment and to some extent, a mind shift. Aspects of this adjustment include (a) realizing that one has a relatively extensive amount of time to observe and reflect, while a child is solving a problem; (b) realizing that it is important to provide adequate wait time for the child; and (c) fully appreciating the importance of observation and reflection on the teachers' part, during the time of waiting. Thus Principle 8 highlights that it is important to engage children in sustained periods of hard thinking. During these periods children do have an opportunity to reflect on their thinking and reflect on the results of thinking (for example, a tentative answer or part solution). It is during periods of sustained thinking that children can reorganize their thinking and change the ways that they not only solve tasks but construe them as well. The scenarios are replete with instances where the teacher provides the child with sufficient time to solve a problem and, consequently, the child is engaged in an extended period of thinking.

Principle 9

Children gain intrinsic satisfaction from their problem-solving, their realization that they are making progress and from the verification methods they develop.

In the individualized teaching sessions that characterize MR there are many instances where children seem to react very positively to challenge. The authors have observed many instances of MR teaching. From these observations it is very clear that having to solve problems that they find difficult does not of itself result in negative attitudes to learning on the part of children. By way of contrast, what seems to be the case is that frequent and ongoing success in solving challenging problems very often results in dramatic improvements in children's attitudes to learning. Related to this, when children are experiencing success in this way, it is not difficult for teachers to find ways to celebrate progress (for example, in discussions with children, parents or class teachers). A related point is that tasks presented in MR sessions can often be organized in ways that lead naturally to children checking or confirming their answers. To sum up, factors such as these seem to be particularly motivational for children – experiencing success with difficult problems, making good and ongoing progress, being aware of and celebrating progress, and checking solutions and developing new ways to confirm answers. There is clear evidence in each of the scenarios that the child gains intrinsic satisfaction from working hard and succeeding at problem-solving. In Scenario 4, for example, on two occasions Joanna directs Wendy not to say the number of bundles in all, that is, Joanna is asking Wendy not to make the task easier.

SECTION B – KEY ELEMENTS OF INDIVIDUALIZED TEACHING

Section B contains descriptions of 12 key elements of MR teaching. Collectively the authors have worked with hundreds of MR teachers and have observed many hours of videotaped records of MR teaching. We believe this set of 12 elements provides a sound basis for critical observation and analysis of this teaching. At the end of this section there is a discussion of an element of teaching labeled 'behavior eliciting' which we regard as inappropriate.

1 Micro-adjusting

The topic of micro-adjusting was discussed briefly in Principle 6 and is explained in more detail here. Micro-adjusting refers to the teacher's role of making fine adjustments to tasks that are presented to a child. These adjustments are made on a moment by moment basis during the course of the child's problem-solving episode. The teacher begins by presenting a task that the teacher hypothesizes is at an appropriate level of difficulty for the child. The teacher observes closely the strategy used by the child and attempts to gauge the difficulty of the task. In the case where the child is unable to solve the task an appropriate micro-adjustment would involve posing a similar but less challenging task. In the case where the child solves the task with some effort micro-adjusting would involve posing a similar but more difficult task. The desired result is that, as a consequence of ongoing micro-adjustment, the child has solved a sequence of tasks of increasing difficulty. Micro-adjustment in this way can lead to development of new strategies by the child. Micro-adjusting can be seen in each of the scenarios and is most evident in Scenario 1. Over the course of this excerpt, Tania solves a sequence of six incrementing tasks of increasing complexity and difficulty, involving bundling sticks (that is, tens and ones). The first three tasks involved displayed materials. These tasks are: (a) 24 and 10; (b) 34 and two 10s; (c) 54 and three 10s. The next two tasks involve screening the initially displayed set of tens and ones: (a) 38 and 10; (b) 48 and two 10s. The final task involves screening both sets of materials – 68 and three 10s. This scenario constitutes an excellent illustration of ongoing micro-adjustment.

2 Scaffolding

Scaffolding (Wood et al., 1976, p. 90) refers to actions on the part of the teacher to provide support for the child's learning in an interactive teaching session. The teacher is engaged in presenting a sequence of related tasks to the child. In doing so the teacher provides assistance to the child for one or more tasks at the start of the sequence, and gradually reduces the level of assistance and withdraws the support as the sequence of tasks progresses. The assistance can take several forms, for example (a) describing or indicating the initial part of the strategy, (b) displaying part or all of a task setting which is otherwise concealed, or (c) demonstrating all or part of a strategy. Scaffolding is evident in Scenarios 1, 3 and 4. In Scenario 3, Jane uses scaffolding on several occasions, for example, when Claire is attempting to solve the missing addend task $2 + x = 4$, Jane says, 'Do you think there's one? Does that make four?' Another example occurs later – on the task $3 + x = 4$, Jane says, 'I've got one, two, three here. How many more to make four?' In Scenario 4, Wendy uses scaffolding when she says 'Four tens?', rather than merely indicating that there is one more ten.

3 Handling an Impasse

Problem-based teaching in individualized teaching sessions involves creating instructional situations in which the child is engaged in sustained periods of hard thinking and learns through solving challenging problems. In such situations, times arise when the child appears unable to solve a particular problem at hand. The teacher in such a situation is likely to try to provide an appropriate micro-adjustment or to provide a scaffold for the child's learning. We use the term 'impasse' to describe a situation where the child is unable to solve the problem at hand despite attempts to micro-adjust or scaffold by the teacher. Typically, these situations are characterized by implicit obligations on the part of the teacher and child, which together lead to a sense of unease. The teacher is uneasy because she has an obligation to provide the necessary micro-adjustment or scaffold, to help the child to solve the problem, and the child is uneasy because he or she has an obligation to use the adjustments or scaffolds to solve the problem. These impasses are usually resolved in one of three ways: (a) the teacher directly releases the child from the obligation to solve the task; (b) the teacher micro-adjusts or provides scaffolding to such an extent that the task is made significantly easier for the child; (c) the child arrives at a means to solve the task. In Scenario 3, on several of the tasks, Jane lifts the screen and so releases Claire from the obligation to solve the task (like (a) above). In Scenario 1 on the last task (68 and three 10s), Tania reaches an impasse when she says 'I don't know what comes after sixty-eight', whereupon Kathryn makes a within task setting change. This is discussed in element 8 below.

4 Introducing a Setting

Mathematics Recovery individualized teaching involves using a range of settings, that is, devices and materials which are used in posing tasks to the child. When introducing a new setting to a child it is important to undertake preliminary explanations and activities in order for the child to become familiar with the setting. The introduction of a new setting usually proceeds as follows. The teacher places the setting on the table and tells the child what it is called. The teacher then proceeds with a series of questions which have the purpose of revealing the child's initial sense of, and ideas about, the setting. In this way the teacher is able to gain insight into the ways in which the child is likely to construe and think about tasks presented using the setting. Introducing a setting in this way has the purpose of making the teacher explicitly aware of the child's ambient thinking about the setting. Related to this, proceeding in this way should reduce the likelihood that the teacher makes unwarranted assumptions about how the child will construe tasks which are presented using the setting, and about how the child will think about elements of the setting when attempting to solve tasks based on the setting. In Scenario 2, when Ivan introduces the ten frame he refers to it as 'this card', and thus does not deny Colin the opportunity of figuring out that it has 10 squares. Ivan questions Colin about what he sees in a very open way and incidentally associates the term 'pairs' with the columns of the frame (containing two squares). This is an excellent example of a low-key and effective way to introduce a setting.

5 Pre-formulating a Task

Pre-formulating a question refers to statements and actions by the teacher, prior to presenting a task to a child, that have the purpose of orienting the child's thinking to the coming task (Cazden, 1986; McMahon, 1998, pp. 21–2). Pre-formulating thus draws the child's attention to key elements relating to the task setting or directs the child's thinking to related tasks solved earlier in the teaching session or in an earlier session. Pre-formulating has the purpose of laying a cognitive basis for a new task or sequence of tasks that the teacher intends to present to the child. In Scenario 2, Ivan pre-formulates when he says, 'Let's see. I'm going to change that again'. In Scenario 3, Jane pre-formulates at the start of the excerpt by saying, 'I am going to put three counters there'.

6 Reformulating a Task

Reformulating a question refers to statements and actions by the teacher after presenting a task to a child, and before the child has solved the task (Cazden, 1986; McMahon, 1998, pp. 21–2). Reformulating has the purpose of refreshing the child's memory of some or all of the details of the task or providing the child with additional information about the task which is thought to be useful to the child. Reformulating is appropriate when the child seems genuinely to be unaware of critical information relating to the task, or to require additional information. Reformulating is inappropriate when the child does not require a restatement of critical information or does not require additional information about the task. Inappropriate reformulation can be distracting to the child and hinder his or her attempt to solve the task. In Scenario 3 Jane appropriately reformulates on several occasions, for example, when she says 'I've only got four counters'.

7 Post-Task Wait-Time

Post-task wait-time refers to the teacher's behavior in providing sufficient time after posing a task for the child to think about and solve the task (Brophy and Good, 1986; McMahon, 1998). In MR individualized teaching, children frequently spend relatively long periods of time (for example, up to one minute or longer) thinking about and solving tasks. Provision of this kind of wait-time is regarded as important for children's learning.

The key to providing sufficient post-task wait-time is to realize that typically during the wait-time the child is engaged in sustained and active thinking. Appropriate wait-time in combination with well-chosen tasks provide the basic ingredients for advancements in and reorganizations of a child's thinking. Some teachers, particularly those who are accustomed to relatively rapid paced interactions with children, may find difficulty in adopting a teaching approach in which the provision of adequate post-task wait-time is an important feature. Thus it is important that teachers are provided with good examples of teaching segments in which the provision of post-task wait-time is highlighted. There are several examples of appropriate post-task wait-time in each of the scenarios. In Scenario 1 for example, there are six occasions, when Tania is apparently engaged in thinking about a task for a period of time exceeding 10 seconds. There are no indications that these periods are stressful or unpleasant for Tania. We believe that provision of wait-time in this way is important for children's learning.

8 Within-Task Setting Change

Within-task setting change refers to a deliberate action on the teacher's part in changing a setting during the period when the child is attempting to solve a task. This often occurs when the child apparently reaches an impasse, that is, the teacher perceives that the child is unable to solve the task that they are currently attempting. In changing the setting the teacher deliberately introduces new elements which, from the teacher's perspective can be linked to elements in the original setting. Thus the intention on the teacher's part is that the new elements enable the child to reconceptualize the current task, and arrive at a solution which was not available to the child before the change of setting. These kinds of changes to a setting during the course of the child's problem-solving can serve to assist the child in arriving at a solution. At the same time, within-task changes of settings can sometimes fail in the goal of leading to a necessary insight or reconceptualization on the part of the child. This often occurs because the child is not able to conceive of the links between the new and the old settings, although these links might be very evident for the teacher. In Scenario 1 Kathryn makes an appropriate within-task setting change when Tania says, 'I don't know what comes after sixty-eight'. This involves producing a hundreds chart. As well as changing the setting, Kathryn provides ample scaffolding for Tania, by counting from sixty-eight to seventy-eight while pointing at the corresponding numeral on the hundreds square.

9 Screening, Color-Coding and Flashing

Screening refers to a technique used in the presentation of tasks in the individualized teaching of MR. A simple example of screening is when an additive task involving two small collections of counters is presented as follows. One collection (for example, the larger) is first displayed and then screened, and the teacher tells the child the number of counters in the collection. This is followed by similarly displaying and then screening the second collection, and then telling the child the number of counters in the second collection. This approach of screening collections during the course of presenting a task to a child was initially used in the case of simple additive tasks and simple subtractive tasks of various kinds (for example, missing addend, missing subtrahend, comparison) involving relatively small numbers. In the course of developing new settings and tasks for individualized teaching in MR, this technique of screening was frequently incorporated into new tasks and settings.

Color-coding is another technique used in the presentation of tasks in MR. In the case of an additive task involving two collections or a missing addend task, counters of contrasting colors (for example, red and green) are used for the collections. In similar vein to screening, color-coding was incorporated into a range of new settings and tasks which were developed for MR.

One benefit of using screening and color-coding in this way is that, during interactive teaching the teacher can direct the child to check their solution. This will involve the child removing the screen or screens and typically using counting to check their solution. In the case of some tasks (for example, addition or missing addend involving two collections), the use of color-coding when posing the task, facilitates a child's checking. Thus in the case of a missing addend task which is presented using a different color for each of the two addends, the child can check his or her solution when the screen is removed by counting the counters in the second collection (that is, the missing addend). The counters in this collection are clearly differentiated by color, from the counters in the other collection (that is, the known addend).

The technique of flashing is used in the presentation of tasks which involve spatial patterns or settings for which spatial arrangement or color-coding is particularly significant. The term 'flash' is used in the sense of display briefly (typically for about half a second). Thus spatial patterns for the numbers from one to six can be flashed and children can be asked to trace the patterns in the air. In similar vein a ten frame with four red dots and six green dots can be flashed and children can be asked to describe the numbers of dots that they saw.

The techniques of screening, color-coding and flashing are used because we believe they are likely to support children's imaging (used in the sense of making a picture in the head) and reflection (used in the sense of thinking about one's thinking). Imaging and reflection we believe are important for cognitive advancement and reorganization of one's current ways of thinking (see the section on child reflection later in this chapter). Use of screening, color-coding and flashing can be seen in many of the teaching procedures in Chapters 5–9. Scenarios 1, 2 and 3 involve significant use of screening by the teacher. In Scenario 3 Jane uses screening and color-coding when presenting missing addend tasks with collections of counters. In Scenario 2 Ivan uses screening, color-coding and flashing in conjunction with each other when using a ten frame.

10 Teacher Reflection

Teacher reflection refers to mental processes undertaken by the teacher during individualized teaching sessions in MR. These mental processes include observation and monitoring of the child's speech and actions, modifying one's view of the child's current knowledge and strategies, and thinking about which tasks should be next presented to the child. During individualized teaching, the teacher, as a general rule, has a clear idea about the likely way that a child will respond to a task or question. Nevertheless, teachers should always closely observe the child's problem-solving and be on the lookout for interesting insights into child's thinking and strategies which might be available to the observant teacher. The teacher's instructional agenda, that is, the questions and tasks which he or she intends to present to the child, should be strongly influenced by his or her ongoing observations and reflections. The agenda should not be predetermined and inflexible, to the extent that the teacher's moves are largely independent of the particular responses of the child. By way of contrast, the reflective teacher has a predetermined instructional agenda, but that agenda is mediated by ongoing observation of and reflection on the child's problem-solving activity. In each of Scenarios 1–4 the teacher is reflecting well on the child's learning. This is evidenced by the highly interactive nature of the teaching.

11 Child Checking

The issue of the child checking his or her solution after solving a task was discussed briefly above with reference to particular circumstances (see 'Screening and Color-Coding'). The general issue of the child checking his or her solution is an important one in the individualized teaching in MR. Through routinely being directed to check their solutions children begin to develop a sense of the notion of verification in mathematics. Verification as used here refers to the general idea that solutions to mathematics problems often lend themselves to being checked or confirmed by a procedure different from that by which the child initially solved the problem. In the case of additive and subtractive tasks, for example, this might involve counting a collection that was not available in the child's visual field when he or she initially solved the task. In the case of other tasks, checking might involve using a device (for example, a hundred chart, a numeral track) that was not available at the time of initially solving the task. Thus the teacher is responsible for triggering the process of checking after the initial attempt to solve the task. This triggering of checking will typically involve asking the child to check his or her response and at the same time making some critical change to the setting (or inducing the child to do so), which essentially recasts the original task in a format which lends itself to checking.

Checking in such cases will typically involve solving the task again, using a less advanced strategy or less advanced thinking. Another important feature of checking of this kind is that the child takes responsibility for confirming his or her solution. There is often an unstated recognition on the part of the teacher and child, that the child has solved the task and confirmed it by checking. As well, checking of this kind can become an established norm of the teaching sessions.

Scenarios 2 and 3 involve several instances of the child checking their solutions. In many cases this is associated with using screening (see element 9 above). The child interprets the teacher's removal of the screen as a signal to check their solution. In Scenario 2, for example, when Ivan removes the screen from the ten frame, Colin spontaneously counts the red counters. Counting the red counters constitutes an appropriate check of the solution provided by Ivan immediately prior to the act of checking. In Colin's MR sessions with Ivan, this is an established norm. In Scenario 4, near the end, when Joanna answers 'Ten bundles', Wendy does not comment on the correctness of Joanna's solution. Rather she directs Joanna to count the bundles. The period immediately before Wendy says 'Well done!' is characterized by a shared but implicit awareness on the part of teacher and child, that Joanna has figured out that if there are 100 sticks then there are 10 bundles of ten.

12 Affirmation

Affirmation relates to statements or actions by the teacher which have the purpose of affirming effort or achievement on the part of the child. In individualized teaching there are likely to be many opportunities for the teacher to assert positively that the child has thought hard to solve a problem. As well, there are likely to be many opportunities for the teacher to speak positively about the child's progress. Our experience has been that teacher affirmation of their effort and achievement is a very positive experience for children and can be done with relatively little time and effort on the part of the teacher. For some teachers, child affirmation seems to fit seamlessly into teaching. As well, affirmation is often quite subtle or tacit but at the same time important and pleasing for the child. Finally, teachers considering how and when to affirm child effort or achievement should realize that affirmation can be rendered ineffective if it takes the form of routinely making a positive comment irrespective of the levels of child effort or achievement. In the course of observing MR teaching sessions, we have seen many instances were appropriate and justified affirmation is used effectively by teachers. In Scenarios 1 and 4 there are a good number of instances of appropriate teacher affirmation. This is not surprising given that in these scenarios the children solve a succession of tasks. In Scenario 2 Ivan affirms appropriately on several occasions, but this is typically non-verbal – Colin looks at Ivan with an expression that seems to say, 'I'm right!' and Ivan responds with an expression that indicates, 'Yes, well done!' In Scenario 3 affirmation occurs at the end of the excerpt, which is entirely appropriate given that this is precisely where Claire uses a strategy that did not seem to be available to her earlier in the session.

Behavior-Eliciting

Behavior-eliciting refers to interactive teaching during which the teacher seems to have a goal of eliciting a particular behavior from the child. The teacher poses a task or a sequence of similar tasks and focusses on having the child respond with a particular behavior that is predetermined by the teacher.

As a general rule this kind of teaching is regarded as inappropriate because of its emphasis on the child behaving in a particular way rather than the child advancing his or her knowledge by developing a new strategy appropriate for the problem. Typically when behavior-eliciting teaching is taking place the child seems to be aware of an obligation to the teacher to behave in a certain way. In these cases the child undergoes a cognitive shift from attempting to solve a problem to attempting to figure out the kind of behavior required by the teacher. As the interaction progresses, the teacher feels obliged to provide more and more clues to the child in order to elicit the required behavior. Because this kind of teaching does not involve active problem-solving on the part of the child, it is unlikely to result in advancements and reorganizations in the child's thinking. Children who are frequently subject to behavior-eliciting teaching are unlikely to experience the benefits of intrinsically motivated problem-solving. The child who is frequently subjected to behavior-eliciting teaching is experiencing teaching in a fundamentally different way from the child who is frequently engaged in intrinsically motivated problem-solving. The latter can result in very positive attitudes to thinking and learning, whereas the former can result in negative attitudes to thinking and learning.

SECTION C – CHARACTERISTICS OF CHILD PROBLEM-SOLVING IN INDIVIDUALIZED TEACHING SESSIONS

Section C contains descriptions of nine characteristics of child problem-solving during MR teaching sessions. As with the key elements of teaching in Section B and the guiding principles in Section A, these characteristics were developed from our observations of many hours of MR teaching. This set of characteristics provides a sound basis for understanding and analyzing children's learning in MR teaching sessions.

1 Cognitive Reorganization

Cognitive reorganization (Steffe and Cobb, 1988, p. 46) refers to a significant change in a child's thinking which typically occurs during the course of attempting to solve a problem. This change involves a qualitative change in the way the child regards the problem and generation of a strategy that was previously unavailable to the child. Cognitive reorganization is often preceded by an extended period of sustained hard thinking on the part of the child. Cognitive reorganizations can be regarded as milestones in the child's learning and development. Near the end of Scenario 3 Claire undergoes a cognitive reorganization, relating to the way she views missing addend tasks. Prior to that point she apparently did not have a strategy to solve these tasks. Her strategy of saying 'Six, seven' while raising two fingers, and then saying 'Two more', indicates that she keeps track of her counts. Prior to this point in the teaching session she had not used this strategy. Near the end of Scenario 4 Joanna undergoes a cognitive reorganization when she figures out that 100 single sticks constitutes 10 bundles of ten. This reorganization is preceded by a period of 22 seconds during which she presumably was trying to figure out how many bundles of ten make 100.

2 Anticipation

Anticipation is a cognitive process that occurs when a child is attempting to solve a problem and typically occurs at the initial phase. Anticipation refers to a realization by the child prior to using a strategy, that the strategy will lead to a particular result (Steffe and Cobb, 1988, p. 231). Anticipation occurs, for example, when a child realizes that, in order to solve a missing addend task, he or she can count-on and, at the same time, keep track of the number of counts he or she makes. It is the simultaneous awareness on the child's part that he or she can both count-on and keep track of his or her counts while doing so that is the essence of anticipation. Near the end of Scenario 3 Claire's cognitive reorganization (see element 1 above) includes anticipating that she can count-on from five to seven and keep track of the number of counts. Anticipation refers to her prior realization that she can keep track. In similar vein, the counting strategies used by Tania (Scenario 1) include the element of anticipation, for example, when she counts-on by tens from fifty-four, she anticipates that she can count-on by tens and stop after she has made three counts.

3 Curtailment

Curtailment is the mental process of cutting short or curtailing an aspect of problem-solving activity (Krutetski, 1976). Prior to commencing the problem-solving activity, the child has an awareness of the results of the activity and thus the activity becomes redundant. Curtailment occurs, for example, when a child realizes that in order to find the number of items in two collections, it is not necessary to count from one. Rather, it is sufficient to count-on from the number of items of the first collection. In Scenario 4 Wendy asks Joanna how many bundles of ten there are, given that there are 80 sticks. Initially, Joanna sets out to count the bundles but Wendy discourages her from doing so. We could say that, at that point Joanna curtails counting from one to figure out how many bundles. She realizes she can work it out and, presumably, in doing so she uses either the number word 'eighty' or the numeral '80', each of which symbolizes how many sticks there are. In similar vein, curtailment occurs when she works out how many bundles in the cases of 90 and 100 sticks.

4 Re-presentation

Re-presentation refers to a particular kind of cognitive activity on the part of the child. This cognitive activity is akin to a mental replay of prior cognitive experience (von Glasersfeld, 1995, pp. 93–4). Thus the child presents again (hence 're-presents') to herself or himself, the prior cognitive experience. Representation in the sense just explained can be viewed as an essential part of the child's problem-solving activity (that is, as far as the child is concerned). In the case of a child who is more advanced in terms of problem-solving capacity, the act of re-presentation would be viewed as unnecessary or redundant. In the case of a child who is less advanced the act of re-presentation is not available to him or her. For this child the actual cognitive experience is required rather than the re-presented experience. Re-presentation in the sense described here can be illustrated by considering the different strategies used by children to solve a simple additive task involving two collections, for example 8 and 4, which are screened. The child for whom re-presentation is unnecessary solves the task by counting on from 8. The child for whom re-presentation is necessary solves the task by counting the first collection from one, and then continuing to count the second collection. For this child it is necessary to replay the cognitive experience of counting the first collection. Finally, the child who is unable to use re-presentation can solve the task only if the collections are not screened. For this child it is necessary to have the experience of actually counting the counters. The differences in the children's abilities to solve the task are attributable to qualitative differences in their concepts of number. Re-presentation is frequently seen in the counting strategies used by children at Stage 2 of the Stages of Early Arithmetical Learning (see Chapter 1), because at Stage 2 children solve tasks involving screened collections but, in doing so, use the redundant activity of counting from one, rather than counting-on.

5 Spontaneity, Robustness and Certitude

These are important characteristics of a child's problem-solving activity. A child's strategy is spontaneous when it arises without assistance from the teacher or without assistance from particular circumstances occurring in the presentation of the problem. A child's strategy is robust when the child is able to use the strategy over a wide range of similar problems (for example, problems involving a range of numbers or problems presented with different settings). Certitude refers to a child's assuredness about the correctness of their solution to a problem. Answering several similar problems in a row with certitude might indicate that the problems are straightforward or routine for the child, and thus continuing with similar problems might not be conducive to advancing the child's knowledge.

In Scenario 1 Tania solves a sequence of tasks. Her solutions seem to indicate a good degree of certitude, notwithstanding that the tasks are increasing in difficulty and complexity (see element 1 in Section B). In similar vein, Joanna in Scenario 4 indicates certitude – she exudes confidence. Near the end of Scenario 3 Claire solves the missing addend task of $5 + x = 7$. Immediately prior to Claire saying 'six, seven, two more', Jane counts the five blue counters. A reasonable hypothesis is that Jane's count from one to five was significant for Claire. Thus Claire's solution was not spontaneous. In Scenario 2 Colin is trying to figure out how many red counters there are on a ten frame if there are two greens and the remainder are red. Colin answers 'eight' and soon after says 'or seven!' If after Colin said 'eight' Ivan had indicated that his answer was correct, it is unlikely that Colin would have then said 'or seven'. Thus we can say that Colin lacked certitude, and that Ivan skillfully tested Colin's certitude. Knowing that Colin lacked certitude is important. Clearly these tasks are problematic for him and this sequence of instruction is within Colin's ZPD (see principle 3 in Section A).

6 Asserting Autonomy

Children in the individualized teaching sessions of MR were sometimes observed to assert their autonomy as problem-solvers. In such situations the child might implore the teacher not to help them or to allow them sufficient time to solve a problem independently. That children who were initially assessed as relatively low-attaining in their number knowledge and subsequently reached a point after which they frequently asserted their autonomy as problem-solvers is a particularly pleasing aspect of Mathematics Recovery. This kind of behavior typically arises after the child has participated in several teaching sessions, and can be attributed to the child's general sense that, during MR teaching sessions, they are thinking hard, successfully solving problems and making progress in the sense that they can now solve problems of various kinds that they were unable to solve at the start of their teaching cycle. Asserting autonomy is overt and strong in Scenario 4. Joanna does this verbally – 'Don't say it for me' and later 'Don't say it' and again 'Don't', and non-verbally – making vigorous pointing actions at Wendy.

7 Child Engagement

The individualized teaching sessions in MR afford many opportunities for children to be actively engaged in mathematical problem-solving. Clearly child engagement of this kind is an important condition for learning. The ideal situation in the teaching sessions is for the child to apply herself or himself directly and with effort when presented with a problem, and to remain engaged in solving the problem for a relatively extended period if necessary. Experience in MR has shown that interactive teaching in which this kind of engagement becomes the norm can be attained with the majority of children. Of course, attainment of this kind of teaching is dependent to a certain extent on factors inherent to the child and over which the teacher has little or no control. It is dependent on sound teaching on the part of the teacher. Thus the teacher's role is to establish an emotionally supportive environment in which hard thinking and persistence on the part of the child is rewarded with successful problem-solving. There are good examples of intense child engagement in all four scenarios.

8 Child Reflection

Child reflection refers to periods of intensive thinking during problem-solving. During these periods the child's thinking can take the form of reflecting on their own prior thinking, that is, the child reflects on the results of their thinking. Thus child reflection is thought of in a Piagetian sense – by reflecting on the results of his or her thinking the child becomes explicitly aware of elements of his or her thinking that were not consciously part of his or her thinking prior to the period of reflection. This in turn can result in cognitive reorganizations and the development of new and distinctive strategies. It is particularly important that teachers engaged in individualized teaching fully recognize the importance of child reflection for the advancement of knowledge and for learning. Periods of child reflection should not be regarded as problematic or unproductive from the point of view of advancing child learning. For some teachers the initial experience of sustained periods of reflection on the child's part seems to invoke concern that the child is not making good use of his or her own or the teacher's time, or that the period of reflection is essentially a negative experience for the child. In reality, there are few indications that MR children are negatively oriented to undertaking sustained periods of reflection. Such periods of reflection are likely to be an important positive factor in advancing the child's knowledge and learning. There are many instances in the scenarios where there are very strong indications that the child is engaged in intensive thinking and reflection (see Principle 8 in Section A).

9 Enjoying the Challenge

There are many instances in the individualized teaching sessions of MR where children seem to revel in challenging problem-solving. What is clear is that virtually all children – including those assessed as low-attaining – can be taught in ways such that, ultimately, they come to regard problem-solving as an intrinsically satisfying and rewarding experience. When this state of affairs is reached in the teaching cycle, learning is regarded as an exciting and positive experience, and important conditions for productive child learning and advancement have been attained. Enjoyment of problem-solving on the part of the child is very evident in Scenarios 1 and 4.

EXAMPLES OF LEARNING AND TEACHING IN INDIVIDUALIZED SESSIONS

We now present four scenarios of Mathematics Recovery teaching taken from individualized teaching sessions. The scenarios begin with an overview and serve to exemplify many of the topics discussed in Sections A, B and C of this chapter. The teaching procedures and settings used in the scenarios can be linked to Key Topics which appear in Chapters 5–9.

Scenario 1 – Kathryn and Tania

Overview

This scenario focuses on incrementing tens and ones by one or more tens (for example, incrementing 54 by three tens), and involves a setting of bundling sticks organized into bundles of ten and singles. Thus Tania's tasks take the form of starting with a given number involving tens and ones and incrementing the number by a given number of tens. In the first part of the scenario Kathryn does not use screens with the sticks, that is, the bundles of tens and ones are displayed rather than screened. Tania's first task is to ascribe number to 2 tens and 4 ones (that is, twenty-four). She then solves the following incrementing tasks: 24 and one ten, 34 and 2 tens, 54 and 3 tens. From this point on Kathryn uses a procedure of displaying and then screening the initial number to be incremented. Tania's next task is to ascribe number to 3 tens and 8 ones. She then solves the following incrementing tasks: 38 and one ten, 48 and 2 tens. At this point Kathryn modifies her procedure by also screening the number of tens by which the number is to be incremented. Kathryn presents the following task: 68 and 3 tens. Tania does not solve this task initially, but ultimately does so with assistance from Kathryn.

Scenario 1

K: (Places out two bundles of ten and four ones to the right of the two bundles.) I'll make a number here and you tell me what it is.

T: Twenty, (points to each of the four ones in turn) twenty-one, twenty-two, twenty-three, twenty-four!

K: There's twenty-four there (places out another bundle of ten to the right of the four ones) and ten more?

T: Twenty-four. (Thinks for 15 seconds while looking at the extra bundle.) Thirty-four!

K: Well done! (Places the three bundles of ten together. Then places out another two bundles to the right of the four ones.) Now, what's thirty-four and twenty more?

T: (Looks at the extra two bundles of ten for 12 seconds.) Fifty-four!

K: Yes! (Places the five bundles of ten together. Then places out another three bundles to the right of the four ones.) Now, what's fifty-four and thirty more?

T: (Touches the three tens, pauses briefly and then points to the five tens and four ones.) Is that fifty-four?

K: (Nods.)

T: (Looks and touches the three tens for 12 seconds and then picks up the three tens.) Now, fifty-four, (places down each of the three bundles in turn) sixty-four, seventy-four, eighty-four!

K: Fantastic! (Places out three bundles of ten and eight ones to the right of the two bundles.) Tell me what that number is?

T: Thirty, (moves each of the ones in turn) thirty-one, … thirty-eight!

K: Good. Thirty-eight, (places a cover over the three tens and eight ones) I'm going to cover that up. How many's under there?

T: Thirty-eight.

K: Thirty-eight, (places out a bundle of ten beside the screen) and ten more?

T: (Picks up the bundle of ten and looks intently at it for 8 seconds.) Hmm, (looks at the bundle for 4 seconds). Forty-eight!

K: (Looks at Tania and smiles.) Well done! (Places the bundle of tens under the screen. Lifts the screen momentarily.) Forty-eight under there (places out two more tens) and twenty more?

T: (Looks intently at both tens for 18 seconds and then looks at Kathryn.) Sixty-eight!

K: Very well done! Excellent work! Okay, let's try two screens. Sixty-eight under there (points to the first screen and then places out three tens under a second screen), and thirty under there?

T: (Looks at the second screen for 12 seconds and then looks at Kathryn.)

K: Sixty-eight and thirty more?

T: (Looks at the second screen for 19 seconds.) I don't know what comes after sixty-eight.

At this point Kathryn placed out a hundreds square, pointed to sixty-eight, and then counted from sixty-eight to seventy-eight while pointing at the corresponding numeral on the hundreds square. Kathryn then pointed in turn to the numerals 78, 88, 98, while Tania said, 'seventy-eight, eighty-eight, ninety-eight'.

Scenario 2 – Ivan and Colin

Overview

In this scenario Ivan introduces the setting of a ten frame. Initially he engages Colin in discussing the frame. He then places 10 red counters on the frame, and poses a sequence of tasks which involve replacing one or more of the red counters with green counters. In doing so he conceals from Colin, his act of replacing one or more red counters. Ivan uses the technique of flashing the ten frame, and Colin's tasks involve determining the number of green counters (for example, one or two) and the number of red counters remaining.

Scenario 2

I: (Places out a ten frame.) Let's have a look at this card.

C: What's that for?

I: (Places 10 red counters on the card.) Well let's work out what we can do with it. What can you see on there, Colin?

C: A pattern.

I: Hmm. What can you tell me about that pattern?

C: They got —. (Looks at the card for 6 seconds.) They got ten.

I: There are ten counters on there.

C: (Nods.)

I: Anything else you can tell me about it?

C: (Waves his hand over the ten frame.) 'Cos they're going down.

I: They're in pairs. (Points to one column of two counters.) One underneath the other like that. And there are ten of them.

C: (Nods.)

I: (Places a screen over the ten frame and replaces a red counter with a green counter without displaying the frame.) Have a look now. I'm only going to show you quickly. (Momentarily unscreens the frame.) What can you tell me about that?

C: One's green.

I: One's green, so how many red ones are there? (Waves his hand over the screen.) How many are there altogether?

C: Ten.

I: There are ten and one of them is green so how many red ones?

C: Nine!

I: (Removes the screen.)

C: (Points to each red counter in turn.) One, two, … nine.

I: Let's see. I'm going to change it again. (Places a screen over the ten frame and replaces another red counter with a green counter to make a column of two greens at the end of the frame). Are you ready to look? (Momentarily unscreens the frame.)

C: Two green!

I: Two greens.

C: And there would be – (looks ahead for three seconds.) eight!

I: (Looks at Colin without indicating whether or not his response is correct.)

C: Or seven!

I: Well, you work it out. Are there eight or are there seven?

C: (After two seconds.) Eight.

I: And why do you think there are eight?

C: 'Cos eight goes before seven.

I: (Unscreens the frame.)

C: (Points to each red counter in turn.) One, two, … eight.

I: Eight reds and two greens. And how many altogether?

C: Ni—, ten!

I: (Places a screen over the ten frame and replaces a red counter with a blue counter making 7 reds, 2 greens and 1 blue). Are you ready to look again? (Momentarily unscreens the frame.)

C: (Makes four pointing actions in the air.) One blue and red and green and green.

I: How many green ones?

C: (Holds up two fingers.) Two.

I: How many blue ones?

C: One.

I: How many red ones? Can you work it out?

C: I don't know. One?

C: Umm (after 3 seconds) seven!

I: (Unscreens the frame.)

C: (Points to each red counter in turn.) One, two, – seven. (Points at the two green counters and then at the blue counter.) And two and one.

I: (Unscreens the frame and replaces the blue counter with a green making 7 red and 3 green counters. Places the screen over the frame.) How many red ones now?

C: (Looks up to his left and makes three points in the air above the screen and then another three points as before while counting subvocally.) Six!

I: How many green ones?

C: (Holds up two fingers.) Two.

I: Are there? (Momentarily unscreens the frame.)

C: Oh, three.

C: So how many red ones?

I: One, two three — (ceases counting). Six! (Makes three points in the air above the screen and then another three points as before while counting softly.) One, two, three, four, five, six.

I: (Unscreens the frame.)

C: (Points at each red counter in turn.) One, two, three, four, five, six, seven.

Scenario 3 – Jane and Claire

Overview

In this scenario Jane presents missing addend tasks to Claire. She presents these by placing out a small collection of green counters (for example, 3), which she does not screen, and placing a second collection (for example, 2) under a screen without displaying them. The task for Claire is to figure out how many counters must be in the screened collection to make a given total. Claire is not successful in solving the first two of these tasks ($3 + x = 5$, $2 + x = 4$). Further, one could say that she does not seem to have a strategy for solving these, and does not seem to conceptualize the tasks in a way that would lead to a solution strategy. The third and fourth tasks: $3 + x = 4$ and $5 + x = 7$, are initially presented in the same way as the first two were. Claire solves these tasks and in both cases is significantly assisted by Jane.

Scenario 3

J: (Places out 3 blue counters.) I am going to put three counters there (places 2 green counters under a screen without displaying them), and I am going to choose some more counters to make five altogether. (Points to the 3 blue counters.) I've got three counters under here. (Points to the screen.) How many counters are there under here to make five altogether?

C: (Immediately.) Four!

J: You think I've got four? (Points to the screen.) If I've got four here would that make five altogether?

C: (Indicates yes.) Hmm, hmm.

J: (Lifts the screen.)

C: Two!

J: (Points to the counters.) You count them.

C: (Counts with the 3 blue counters and then the 2 green counters.) One, two, three, four, five!

J: (Places out 2 blue counters.) This time I've got two. Think really carefully.

C: This is going to be hard.

J: (Places 2 green counters under a screen without displaying them.) And I'm going to choose some of my green ones, and I've only got four counters altogether. (Points to the 2 blue counters.) I've got two here, and I've only got four altogether. (Points to the screen.) How many are under here?

C: (Looks at Jane and smiles.) Five.

J: I've only got four counters. (Points to each of the two blue counters in turn.) One, two, and some more here to make four.

C: Hmm. (After 3 seconds.) Five.

J: I don't think it's five. Have another think.

C: (Immediately.) Six.

J: Look at them and think. (Puts her hand on the screen.) How many do you think are under here?

C: (Immediately.) Four.

J: (Moves her hand over the screen and the unscreened counters.) I've only got four altogether, (points to the 2 unscreened counters) and there are two here. (Points to the screen.) How many more must be under there to make four altogether?

C: (Looks up and to her left for 10 seconds.) Five.

J: (Moves the screen and displays one of the 2 green counters.) Do you think there's one? Does that make four? (Points in turn to each of the two blue counters and then the green counter.) Would that make four?

C: Hmm. (Shakes her head.) No.

J: (Points to the green counter.) So was there one under there?

C: (Shakes her head.)

J: (Moves the screen and displays both green counters.) Were there two under there? Would that make four?

C: (Looks at Jane and smiles.)

J: (Points in turn to each of the two blue counters and then each of the green counters.)

C: (Immediately.) Yes.

J: And how many were under there?

C: Two!

J: Yes, to make four. What about this one. (Places out 3 blue counters.) Three blue counters there.

C: (Looks at the 3 blue counters.) Yes.

J: (Places 1 green counter under a screen without displaying it.) Think hard. This time I've got some under here and I've got four again. (Points to the 3 blue counters.) I've got three here. (Points to the screen.) How many under here to make four altogether?

C: (Looks ahead for 6 seconds.) Six.

J: Now think hard.

C: (Immediately.) Nine.

J: (Places her hand on the screen and moves it slightly.) If I slid this off how many would I see to make four altogether?

C: (Immediately.) Three.

J: (Moves the screen slightly again.) I'd see another three to make four? (Points to the 3 blue counters in turn.) I've got one, two, three here. (Points to the screen.) How many more would I need to make four? (Points to the 3 blue counters in turn while counting softly.) One, two, three.

C: (Looks straight ahead.) One more.

J: (Unscreens the green counter.) Well done! (Places out 5 blue counters.) I got five here this time.

C: (Immediately, pointing to each counter in turn and counting quickly.) One, two, three, four, five.

J: (Places 2 green counters under a screen without displaying them.) And I'm going to choose some of these and I want to make seven.

C: (Immediately.) Seven.

J: (Points to each blue counter in turn.) One, two, three, four, five. (Points to the screen.) And some more under here to make seven?

C: (Immediately.) Six.

J: Now have a think. You thought hard about the last one. (Points to each blue counter in turn.) One, two, three, four, five. (Points to the screen.) How many would I show under here to make seven?

C: (Immediately.) Six.

J: (Places her hand on the screen and moves it slightly.) If I slid this off how many would I see? (Points to each blue counter in turn.) One, two, three, four, five? (Points to the screen.)

C: One more?

J: One more? How many would that be if there was one more. (Points to each blue counter in turn.) One, two, three, four,

C: (Interrupting.) Five, six!

J: Yes but I want seven!

C: Seven!

J: (Points to each blue counter in turn.) One, two, three, four, five.

C: (Looks ahead for 2 seconds.) Six, seven. (Holds up 2 fingers.) Two more!

J: (Uncovers the screen.) Well done!

Scenario 4 – Wendy and Joanna

Overview

In this scenario Wendy uses a setting of bundling sticks organized into bundles of tens, and a pile of plastic digits which Joanna uses to make decade numbers (for example, a '4' and a '0' is used to make 40). Wendy presents a progression of tasks involving adding a bundle of ten for each subsequent task. Joanna has the task of saying the decade number corresponding to the number of sticks and making the number from the plastic digits. The scenario begins with Joanna working out that two bundles contain 20 sticks and building '20'. As the scenario progresses Joanna asks Wendy not to tell her the number of tens in all, that is, each time she places out an extra bundle. Apparently Wendy wants to work out for herself that, given seven bundles, one extra bundle makes eight in all, as well as working out that there are 80 sticks in all. Joanna's last task, that of determining how many bundles correspond to 100 sticks, proves to be challenging for her.

Scenario 4

W: (Gives Joanna two bundles of ten.) We have one ten and another ten, so how many altogether?

J: (Looks at the two bundles.) One hundred!

W: Try again! Think! You're not allowed to count them. Ten there, and another ten. What would ten and another ten be?

J: Twenty.

W: Well done! Make a twenty.

J: (Looks at the pile of plastic digits. Places out a '2' and then a '0'.)

W: Well done. (Places out a third bundle.) Well what would we have if I put another ten there?

J: Thirty.

W: Make it!

J: Let me do it fast. (Quickly places out plastic digits to make '30'.)

W: (Places out a fourth bundle.) Four tens?

J: (Quickly places out plastic digits to make '40'.)

W: How many is that?

J: Forty.

W: (Places out a fifth bundle.) Five tens?

J: (Quickly places out plastic digits to make '50'.) Fifty.

W: (Places out a sixth bundle.) Six tens?

J: (Quickly places out plastic digits to make '60'.) Sixteen.

W: Pardon?

J: Six-ty (emphasizes 'ty'.)

W: (Places out a seventh bundle.) Seven tens?

J: (Makes vigorous pointing actions towards Wendy.) Don't say it for me. (Quickly places out plastic digits to make '70'.) Seven-ty (emphasizes 'ty').

W: Got it! Well done! (Picks up another bundle.)

J: Don't say it.

W: All right, here is another ten. (Points to the eight bundles.) How many tens are there now?

J: (Waves her hand at Wendy.) Don't – (Places out plastic digits to make '80'.) Eighty!

W: (Points to the eight bundles.) Yes there are eighty sticks but how many bundles are there?

J: (Looks uncertainly at the plastic digits making '80', moves the digits and begins to pick them up.)

W: Yes. (Makes '80' again with the digits.) That's right. You've done the right thing. I'm asking you a different question.

J: Eight-ty (emphasizes 'ty').

W: That's true. Yes, there are eighty sticks but —?

J: (Points to the first bundle and begins counting subvocally.)

W: (Quickly places her hand over the eight bundles.) No. How many bundles are there?

J: (Ceases counting and looks at the bundles for two seconds.) Eight.

W: Well done! (Places out a ninth bundle.) Another bundle!

J: (Waves her finger at Wendy.) Don't say —. (Places out plastic digits to make '90'.) Ninety.

W: How many bundles?

J: Nine-ty (emphasizes 'ty').

W: Ninety sticks. Yes

J: Nine!

W: Well done! (Places out a tenth bundle.) Here's another one.

J: Don't tell me. (Looks at the plastic digits for two seconds and then looks at Wendy.) This is hard. (Waves her finger at Wendy. Looks at the plastic digits for three seconds and then begins to pick up digits.) Uh huh, don't tell me. This, this and this (picks up three digits and makes '100', and then looks at Wendy).

W: (Feigns surprise.) Oh! What does that say?

J: One hundred and – (places her hand over her mouth).

W: What does it say?

J: One hundred.

W: (Feigns surprise.) One hundred sticks! There really are one hundred sticks. How many bundles are there?

J: (Looks at the plastic digits showing '100'.) One.

W: One bundle? (Picks up one of the bundles.) No, there's one bundle. (Waves her hand along the line of ten bundles.) How many bundles?

J: (Puts her hands over her face.) I can't

W: (Quickly places her arm over the ten bundles.) Work it out! What would it be?

J: (Puts her hands over her face again and thinks for 22 seconds. Looks at Wendy.) I can't do this one.

W: Yes you can.

J: (Moves the right hand '0' in '100' away, and then looks at Wendy and smiles.) Ten (speaking softly).

W: Pardon?

J: Ten.

W: Ten what?

J: Ten bundles.

W: (Takes her arm off the ten bundles.) Count them.

J: (Pointing to each bundle in turn.) One, two, … ten.

W: Well done!

THE GUIDING PRINCIPLES, KEY ELEMENTS AND CHARACTERISTICS OF PROBLEM-SOLVING REVISITED

During the course of working with MR teachers in several countries, we have found that not only do the guiding principles, key elements and characteristics of problem-solving help teachers to plan, prepare and implement teaching, but they also provide a valuable tool for reflection and professional development. By this we refer to the fact that they can be used in a very positive way to reflect on and evaluate performance. This is useful for teachers working together in the same school but also for teachers working alone.

First, the principles, key elements and characteristics can be used as an aide-mémoire, or checklist, to see if some areas are being over- or under-emphasized during teaching. Second, specific guiding principles, key elements and characteristics can provide guidance at particular stages. This is indicated in Table 2.2 for (a) the planning stage, (b) during a lesson, and (c) after a lesson or a series of lessons. Thus the guiding principles (Section A) (excepting A6), as well as guiding teaching, can guide preparation for teaching. The key elements of MR teaching (Section B) (excepting B10) can guide teaching.

The characteristics of children's problem-solving (Section C) can guide reflection over the course of several lessons. For example, one could ask whether the child is bringing about cognitive reorganization in their thinking. Does the child show an assuredness about solving the problem and know that they are correct without teacher support? If so, we can note the child's certitude. Is the child engaging with the tasks not only because they are novel and interesting, but because a climate has been built in which the child knows they can try solutions without being penalized and that effort and hard thinking have their own rewards? Most of all, is the child enjoying the challenge of problem-solving and is the teacher also enjoying the responsibility of providing that challenge?

We do not advocate using the guidelines in a mechanistic way but there is value in using them as a reflective tool. A very positive aspect is that following a successful lesson the guidelines can be used to determine which elements were useful in bringing progress. On some occasions, reflective thought moves us to a negative stance and it is sometimes necessary to view the positives in order to keep perspectives in balance.

Table 2.2 Using the guiding principles of teaching, key elements of teaching and characteristics of children's problem-solving for reflection

At the planning stage

A1	Problem-based/inquiry-based teaching
A2	Initial and ongoing assessment
A3	Teaching just beyond the cutting edge (ZPD)
A4	Selecting from a bank of teaching procedures
A5	Engender more sophisticated strategies
A7	Incorporating symbolizing and notating
A8	Sustained thinking and reflection
A9	Child intrinsic satisfaction

During a lesson

A6	Observing the child and fine-tuning teaching
B1	Micro-adjusting
B2	Scaffolding
B3	Handling an impasse
B4	Introducing a setting
B5	Pre-formulating a task
B6	Reformulating a task
B7	Post-task wait-time
B8	Within-task setting change
B9	Screening, color-coding and flashing
B11	Child checking
B12	Affirmation

After a lesson or series of lessons

B10	Teacher reflection
C1	Cognitive reorganization
C2	Anticipation
C3	Curtailment
C4	Re-presentation
C5	Spontaneity, robustness and certitude
C6	Asserting autonomy
C7	Child engagement
C8	Child reflection
C9	Enjoying the challenge

3
Whole-Class Teaching

In this chapter we first discuss the two instructional goals of sense-making and intellectual autonomy that we believe underpin whole-class teaching in mathematics. Following this we outline the teaching and learning cycle for whole-class teaching that consists of four parts: (1) Where are they now? (2) Where do you want them to be? (3) How will they get there? and (4) How will you know when they've arrived? The focus of this chapter is teaching whole classes. Nevertheless, many of the ideas in this chapter apply to teaching groups of students as well.

INSTRUCTIONAL GOALS OF WHOLE-CLASS TEACHING

The approach to whole-class teaching that we advocate has two fundamental instructional goals: sense-making and intellectual autonomy. The guiding principles and key elements of teaching, together with the characteristics of children's problem-solving set out in Chapter 2, can also be applied to whole-class teaching and can support the achievement of these fundamental goals.

Sense-Making

Much of the mathematics taught in classrooms is carried out in what is called a transmission mode. In this mode the expert teacher transmits knowledge to the learner. 'Most education systems in every culture operate from this perspective' (Wheatley and Bebout, 1990, p. 107). In this mode teaching is analogous to filling empty vessels. Our view is that this is not an appropriate mode for mathematics teaching today. According to the constructivist view of knowing and learning (von Glasersfeld, 1995) that underlies our approach to teaching, mathematical knowledge cannot be passed on to children. '(L)earning mathematics is an active, problem-solving process' (Yackel et al., 1990, p. 12) where 'each child has to construct his or her own mathematical knowledge … [and children] develop mathematical concepts as they engage in mathematical activity, including trying to make sense of methods and explanations they see and hear from others' (ibid., p. 13).

As indicated above, teaching in the transmission mode seems to involve a belief that mathematical knowledge can be directly transmitted to children. Teaching in the transmission mode also seems to emphasize the teaching of procedures at the expense of helping children to make sense of mathematics. We believe that the teaching of procedures is important. However, procedures should arise out of the sense-making activity of children during mathematics teaching. We contend that the role of the teacher in inquiry-based whole-class teaching is to create a rich and stimulating environment where children are encouraged to generate, reflect on and debate solution strategies, and that these processes should constitute normal practice in mathematics lessons. The implications of this for teaching are discussed later in this chapter.

Intellectual Autonomy

In mathematics lessons, it is common for children to regard as part of the teacher's role the making of decisions about the appropriateness of a response, whether the response is verbal or written. Given teachers' maturity in both age and mathematical experience, one could argue that it is entirely appropriate for them to be the arbiters of children's mathematical responses. Clearly, teachers are continually making judgments about their children's work. However, in our view, it is how teachers communicate these judgments that is critical. The manner and the substance of teachers' responses will promote in children either teacher dependence or an ability to think and function independently. In our view, fostering children's confident and self-reliant use of strategies is an important aim in the teaching of early number and, indeed, more generally in the teaching of mathematics. Thus, in the teaching of mathematics we are advocating a move away from 'teacher-dominated instruction, in which children listen to teacher explanations, respond to teacher directives, and develop expertise using pre-given solution procedures' (Yackel, 1995, p. 131) to an approach where 'children actively engage in doing and talking about mathematics' (ibid.).

In our experience, much of mathematics teaching involves the teacher telling the children what to do and how to do it. As well, it is common to have external factors influencing our teaching, such as tests and examinations that can constitute systems of reward and punishment. In contrast, we are advocating an approach where children routinely are given opportunities to actively solve problems, to reflect on their solution methods and discuss and debate their strategies with classmates and the teacher. In such a setting, children 'develop beliefs about mathematics and about their own and the teacher's role' (Yackel et al., 1990, p. 20) that foster intellectual autonomy. 'Children who are discouraged from thinking autonomously will construct less knowledge than those who are mentally active and confident' (Kamii, 1985, p. 46).

Let us illustrate the point by presenting a scenario that demonstrates how this stance translates to the classroom.

Classroom Scenario 3.1

The focus of this lesson is addition involving a 2-digit number and a 1-digit number. The children are seated in a circle away from their desks. The teacher places two tens (egg cartons modified to each have ten sections) and eight ones (counters) in the circle (see Figure 3.1).

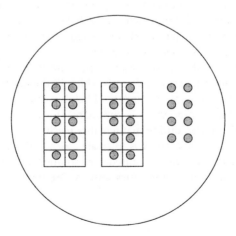

Figure 3.1 Addition task

Teacher: How many counters are there? (Children put up their hands and the teacher points to Jill.)
Jill: Twenty-seven.
Teacher: How do you know there are twenty-seven counters?
Jill: (Points to the tens.) There's ten and ten. That's twenty. (Points to the eight.) I can see five, six, seven, eight. Oh, twenty-eight.
Brad: I can see (points to the eight counters) there are four and four which makes eight. I didn't need to count them!
Teacher: Thanks Brad. That was a quick way to find the eight counters. (Covers the twenty-eight counters with a large box and places another five counters next to the box. See Figure 3.2.)
Teacher: How many counters are there now? (Allows sufficient time for most children to get an answer.) Who can tell us how they worked it out? (Points to Mark.)
Mark: Thirty-three.
Teacher: How did you get your answer?
Mark: I said twenty-eight and then twenty-nine, thirty, thirty-one, thirty-two, thirty-three.
Teacher: Thanks Mark. You explained that very well. Who did it a different way? (She points to Jade.)
Jade: See the five? I used two of them to make thirty and then used the other three to make thirty-three.
Teacher: That's a great way to do it Jade. Tell us again please? Listen carefully to how Jade worked it out. (Jade repeats her strategy to the class.)

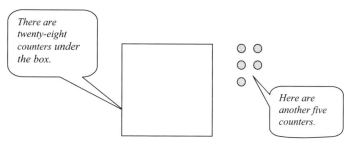

Figure 3.2 Hidden addition task

The two key goals of sense-making and intellectual autonomy are clearly seen in the Classroom Scenario 3.1 (see above). The teacher demonstrated four important actions: (a) posed a meaningful problem; (b) valued the children's responses; (c) asked the children to explain their strategies rather than arbitrate on the correctness of answers; and (d) used appropriate materials to enhance learning.

In this book we are not advocating an approach to instruction based mainly on discovery learning, where children invent mathematics with little direction or guidance from the teacher. However, in the approach we advocate, children's discovery, or construction of knowledge and strategies, is very important. Discovery learning may encompass a range of approaches. If discovery learning involves basing instruction in early number on the belief that children will discover mathematics by interacting with materials (see Cobb, 1991, for a critique of this approach), then discovery learning is inadequate in our view. Discovery learning in that sense seems to involve an assumption that the mathematics is embodied in the materials and that mathematical knowledge will be transmitted to the children who interact with the material. There are occasions we believe, where it might well be profitable for children to play with or engage with materials in an unstructured way. An example is when particular materials are introduced. It is often useful to set aside time for children to interact with new materials before the materials are used as part of a more formal lesson.

pproach we are advocating, the teacher creates a rich mathematical environment in their
where children's mathematical knowledge and strategies can develop. The role of the
crucial in both planning instruction and teaching. The teacher's role is mainly that of expert
guide rather than judge or referee. As expert guide, the teacher plans the mathematical environment
and leads the children on their mathematical journey. This chapter, therefore, has as its focus the role
of the teacher as an expert guide for mathematics learning.

THE TEACHING AND LEARNING CYCLE

In the remaining sections of this chapter we describe what we refer to as the 'Teaching and Learning
Cycle'. This cycle has four key elements.

- ▶ Where are the pupils now?
- ▶ Where do I want them to be?
- ▶ How will they get there?
- ▶ How will I know when they get there?

The Teaching and Learning Cycle been widely used in many of the schools we have worked with as a
basis for planning instruction in early number. As shown in Figure 3.3, the cycle has four elements. In
instruction in early number, each of these elements is critical in long-term planning (for example, for
a school term, or for a unit of work) and in planning specific lessons.

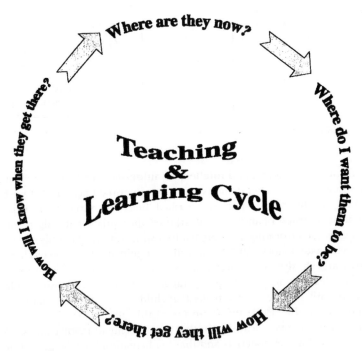

Figure 3.3 The Teaching and Learning Cycle

WHERE ARE THEY NOW?

Before teaching commences, we believe it is essential for the teacher to determine the extent of each child's knowledge. This seems patently obvious. Nevertheless, in our experience it is common for teachers not to give due credence to assessing where their children are in terms of their mathematics learning. Initial assessment is the starting point in the teaching and learning cycle.

Initial Assessment

For whole-class mathematics teaching, the initial assessment is driven by the following questions. First, what aspects of early number should be assessed? Second, how should these aspects be assessed?

The Learning Framework in Number which is described in Chapter 1, provides a comprehensive answer to the first question. In our view it is essential to assess children's knowledge on the following aspects of LFIN:

▶ Aspect Al – the Stages of Early Arithmetical Learning;
▶ Aspects B1 – Forward Number Word Sequences;
▶ Aspects B2 – Backward Number Word Sequences;
▶ Aspect B3 – Numeral Identification.
▶ Children who are at Stage 3 or higher on SEAL, should be assessed on Aspect A2 – Base-Ten Arithmetical Strategies and Part D – Early Multiplication and Division.

Chapter 1 provides a detailed overview of these aspects of early number.

Our answer to the second question regarding the nature of the assessment above involves the use of a distinctive approach to assessment in early number. This approach is interview based and involves presenting a schedule of tasks to the child. The approach has the purpose of providing detailed information about the child's current knowledge and strategies in early number, and in particular the stage and levels of the child's knowledge in terms of the tabulated models of the aspects stated above. These tabulated models appear in Chapter 1. This approach to assessment generates an Assessment Record for each child that includes summary data consisting of the child's stage and levels. Summary data for all of the children in the class is collated using a Class Summary Sheet. This highlights areas of strengths and weaknesses at the class level and thus provides important information for planning. Children's Assessment Records and the Class Summary Sheet are important for documenting children's progress over the course of the school year. This interview-based approach to assessment is described in detail in the revised edition of *Early Numeracy: Assessment for Teaching and Intervention* (Wright et al., 2006).

WHERE DO I WANT THEM TO BE?

This part of the Teaching and Learning Cycle concerns the need on the part of teachers to clarify the objectives of their teaching. In order to do this teachers require (a) sound mathematical knowledge, and (b) knowledge of how children's learning of mathematics progresses.

Teachers' Mathematical Knowledge

We believe it is important that teachers have sound knowledge of mathematical content. Further, we believe teachers who advance their knowledge of the mathematics content that they teach will find this beneficial for their teaching (Ma, 1999). It is beyond the scope of this book to outline how teachers' mathematical knowledge can be advanced apart from encouraging teachers to examine carefully the mathematics they are responsible for teaching and to note any aspects where they believe their knowledge should be advanced. Having done this, they should discuss this issue with colleagues who have the knowledge to assist them or seek support from school administrators who might be in a position to organize appropriate professional development. This might consist of several workshops run by a mathematically competent teacher or consultant. Indeed, it may be appropriate for school administrators to take a lead in this issue and actively seek to raise the mathematical knowledge of their teachers.

One means by which teachers can advance their knowledge of mathematics content relates to the approach to assessment and teaching that is described in this book and the authors' previous book (*ENAT1*) (Wright et al., 2006). Our experience, and that of many of the teachers with whom we have worked over several years, is that the strong focus on studying and documenting children's strategies in numerical contexts that is a feature of the approach described in this book does serve to advance teachers' knowledge of mathematical content. We do not claim that learning this approach will fully meet the needs of teachers who need to learn more mathematics. However, we believe that learning this approach can result in a significant amount of new learning of mathematics for many teachers.

Teacher Knowledge of How Children's Learning of Mathematics Progresses

The second aspect in this 'Where do you want them to be?' section, relates to how children's learning of mathematics progresses. This topic is the focus of Chapter 1. In particular, the LFIN in Chapter 1 provides an answer to this question. Our experience of working with many teachers in a range of school systems is that when teachers begin to learn about the LFIN, they immediately apply this new knowledge to their planning and teaching. Often this involves thinking critically about the number topics they are currently teaching, and considering their suitability, given the current levels of knowledge of their children.

HOW WILL THEY GET THERE?

The third part of the teaching and learning cycle focusses on effective whole-class teaching. Four principles of planning are outlined and each is exemplified in a classroom scenario. Following this, terms such as 'setting', 'activity' and 'task' are explained, and key attributes of inquiry-based, whole-class teaching are described. Three approaches to whole-class teaching (short sessions, problem-solving sessions and special activity sessions) are described. Finally, a lesson that incorporates these three approaches is described.

Planning for Effective Teaching

Classroom Scenario 3.2

Teacher: I have nine oranges. (She puts them in a box.) Here are another three oranges. (These are placed next to the box.) How many oranges are there altogether?

Billie: There are twelve. If you put one more into the box you'll have ten and ten and two is twelve.

Teacher: Thanks, Billie. Are there any other ways of working it out? What did you do, Kim?

Kim: I made nine on my fingers then counted ten and I used my toes, – eleven, twelve. There were twelve oranges.

Teacher: You did really well to tell us that. Does anyone else have any different ideas?

Chris: I didn't run out of fingers. I just kept nine in my head and said ten, eleven, twelve.

Teacher: Thanks, Chris. That's a quick way to work it out.

In Classroom Scenario 3.2 it can be seen that there is a wide range of knowledge among the children. Billie used knowledge of tens and ones to solve the task, Kim made replacements for the items being counted and Chris used a counting-on strategy. This range of knowledge is typical of many classes. The children have a range of levels of number knowledge and strategies. Some children use relatively advanced strategies, while others use less advanced strategies. The following principles of planning for whole-class teaching take account of the likelihood of a range of levels of knowledge among the children.

▶ Teachers need a detailed profile of the current number knowledge of each of their children. (*Where are they now?*)

▶ Teachers must have sound knowledge of the essential elements that advance children's number knowledge. (*Where do I want them to be?*) This will assist teachers to plan a balanced instructional programme which includes activities encompassing all aspects of the LFIN.

▶ Lesson planning must allow for the range of levels of children's knowledge in the class. This involves using flexible grouping in order to meet the individual needs of children. It also involves developing an approach to whole-class teaching that takes account of the range of levels of children's knowledge.

▶ The lessons should be problem-centered where the problems can be solved using a range of strategies characteristic of differing levels of mathematical knowledge.

Settings, Activities, Tasks and Games

In this section we provide descriptions and explanations of key terms such as setting, activity, tasks and games.

Setting

A setting is a physical situation used by the teacher to provide mathematical learning experiences which typically involve children undertaking activities.

Activity

The term 'activity' usually refers to a task or game that has been designed to promote learning in mathematics lessons.

Task

A task is a problem or question posed to the whole class, to a small group or to an individual child. Teachers can use tasks to make judgments about children's number knowledge during lessons. Some tasks invoke a response that involves generating or using a strategy. These are called problem-centered tasks. Other tasks focus on specific items of knowledge (for example, identifying numerals or saying number word sequences). For a given child, a task is problem-centered if the child is unlikely to give an immediate response. It is not essential for problem-centered tasks to be word problems or based on a story situation, that is, it is not essential for problem-centered tasks to be contextual problems. Finally, problem-centered tasks should (a) be locatable in the LFIN, (b) have a clear purpose, (c) promote thinking rather than an automatic response and (d) be challenging for children at a range of levels.

Games

Typically, games are used to supplement tasks that have been posed during whole-class teaching. Games should (a) have a clear purpose, (b) be locatable in the LFIN, (c) have rules that are simple and easy to follow, and (d) be engaging and interesting. Self-checking games release the teacher from close supervision. Having an adult supervisor participating in the game can result in more effective learning, for example, the adult can lead the children in discussing number ideas arising from the game. Games also have other benefits.

Games can advance specific aspects of children's number knowledge. Board games for example, can focus children's attention on the spatial patterns on the faces of a die or focus children's attention on a number word sequence. Score sheets for some games focus on tens and ones. As these examples indicate, a well-constructed game can often advance several aspects of children's number knowledge.

Pair games

Key Attributes of Problem-Centered Whole-Class Teaching

In this part we set out key attributes of problem-centered whole-class teaching. The attributes are organized under the headings of 'Problem-Centered Tasks', 'Classroom Discourse', 'Role of the Teacher' and 'Children in a Problem-Centered Class'.

Problem-Centered Tasks

In the previous section we explained the term 'problem-centered tasks'. In the teaching approach we advocate, use of problem-centered tasks is important. Thus in whole-class teaching, teachers use problem-centred tasks that are designed to engage all the children. The expectation of the teacher is that the tasks will be solved in various ways and by children at various levels of number knowledge. In accordance with recommendations by Perry and Conroy (1994), we advocate using only a small number (no more than three) of problems in each lesson. These problems should be carefully planned. Japanese mathematics classes typically involve one rich problem. This is described by Stigler and

Hiebert (1999), in their report on an international study of mathematical teaching practices, that was part of the Third International Mathematics and Science Study (TIMSS) (for example, ibid., pp. 76–83).

Classroom Discourse

Class discussion   An essential aspect of an effective and mathematically rich learning environment is that children routinely verbalize their thinking and the strategies they use to solve problems. The teacher should promote comment and discussion in response to children's descriptions of their solutions to the whole class (or group). Thus an important role for the teacher is to promote children explaining solutions to the class and discussion of these explanations. Explanation and discussion of solutions under the guidance of the teacher become a routine part of whole-class teaching (Cobb et al., 1997a).

Valuing children's responses   When teachers show genuine interest in children's explanations, they indicate to the class that their ideas are valued. Within such a supportive environment, children are encouraged to think and to attempt to verbalize this thinking. Listening, questioning and debating become established as a normal occurrence in mathematics lessons. 'When children are given opportunities to talk about their mathematical understandings, problems of genuine communication arise. These problems, as well as the mathematical tasks themselves, constitute occasions for learning mathematics' (Yackel et al., 1990, p. 12).

Notation and symbolizing to support children's learning   As described above, an important role for the teacher in problem-centered whole-class teaching is to promote explanation and discussion of children's solutions. One important aspect of the teacher's role in this endeavor is the use of notation and symbolizing as an explicit means to support children's thinking, discussion and mathematical development. The term 'notation' is used here in the sense of writing in a numerical context, and 'symbolizing' is used in reference to arithmetical symbols and includes formal mathematical symbols and informal symbols which might be used by the teacher or children in their discussions of solutions. Our incorporation of notation and symbolizing into problem-centered teaching draws on the work of Cobb and colleagues (for example, Cobb et al., 1995; 1997b; 2000). As well as verbally redescribing children's strategies the teacher uses notation to record their strategies. In this way notation is introduced in a natural way by building on the children's thinking. In doing this the teacher is not limited to conventional notation. From the children's perspective, the notation is introduced incidentally rather than formally. The teacher encourages the children to use the notation during individual activity as well as when describing their strategy to the class. Thus the notation becomes an explicit topic of conversation for the children. As reported by Cobb and colleagues, teachers' proactive development of notation can provide critical support for children's mathematical development. 'These records helped them distance themselves from their ongoing activity and thus reflect on what they were doing. Further, the use of notation contributed to the productiveness of whole-class discussions by helping to make individual children's contributions explicit topics of conversation that could be compared and contrasted' (Cobb et al., 1995, pp. 10–11). The use of notation and symbolizing in this way gradually evolves to more standard forms. This approach to notation and symbolizing can provide for children an important means of mathematization, which is explained in the next paragraph.

Horizontal and vertical mathematization   An important role for the teacher in a problem-centered classroom is to create situations where children can reason mathematically. In these situations the goal is for children to reinvent mathematics through a process which is referred to as 'progressive mathematization' (Gravemeijer, 1994; Treffers and Beishuizen, 1999). Progressive mathematization occurs in two interrelated forms: the first is that children reason initially in a contextual situation, and then the focus of their reasoning shifts to a mathematical situation that in some sense arises from the contextual situation. Thus when mathematization involves a shift in the focus of attention from a contextual situation to a mathematical situation it is referred to as horizontal mathematization. Alternatively, children can reason mathematically in situations which involve contextual problems. In these situations children are reasoning about mathematical ideas which are not directly linked to real-world (or contextual) problems. When reasoning of this kind involves the development of knowledge and strategies which, from a mathematical perspective, are more advanced than the knowledge and strategies initially used in the mathematical situation, this is referred to as vertical mathematization. Thus horizontal mathematization involves 'the analysing of real-world problems in a mathematical way' (Treffers and Beishuizen, 1999, p. 32), whereas vertical mathematization involves 'development of strategies and concepts in a certain area of the mathematics system itself' (ibid.).

Role of the Teacher

Guiding children's mathematical development   'The teacher plays a crucial role by guiding the development of ... a problem-solving atmosphere, an environment in which children feel free to talk about their mathematics' (Yackel et al., 1990, p. 20). The metaphor of the teacher as guide rather than judge who tells children when they are right or wrong, accurately describes our view of the teacher's role during mathematics lessons.

Developing effective whole-class socio-mathematical norms   The teacher is responsible for establishing the socio-mathematical norms (Yackel and Cobb, 1996) for whole-class teaching. These norms relate to the social climate, the commonly agreed modes of operation and the ways in which mathematics lessons proceed. For any teacher and class, socio-mathematical norms are developed both overtly and covertly. Examples of norms that might operate during mathematics lessons are that children individually do exercises from a textbook without discussion with other children and that the teacher routinely explains how each task is to be done and the children perform similar tasks step by step. The norms that are indicative of a problem-centered classroom include valuing children's responses; promoting reflection and discussion; emphasizing problems and thinking, rather than procedural mathematics; and 'that cooperation and negotiation are valued over competition and conflict' (Yackel et al., 1990, p. 20).

Representative of the mathematical community   The teacher is involved in the process of transferring what society deems to be important and relevant through the mathematical content and the way mathematics is taught in each lesson.

Giving appropriate reactions   As stated previously, the teacher values children's responses and attempts by questioning and dialogue to assist children to clarify their thinking. The teacher regards all of the children's responses as indicative of their thinking. The teacher will also legitimize some strategies while de-emphasizing other aspects of children's thinking or mathematical activity (Yackel and Cobb, 1996). This is always done in a way that does not devalue children's efforts to solve problems. The teacher's purpose is to develop children's mathematical knowledge and, in particular, the mathematical sophistication of their responses.

Meaningful activity   There is an emphasis on children engaging in meaningful activity rather than an emphasis on achieving the correct answer. Incorrect answers are regarded by the teacher as indicators of children's thinking and therefore as potentially valuable information for the teacher.

Micro-adjusting   When teachers have a sound knowledge of the LFIN and listen carefully to children's responses, they frequently make minor modifications to questions or activities during a lesson. The purpose of this micro-adjustment is to keep the focus of the lesson at the cutting edge of the children's learning. This fine-tuning becomes a natural part of teaching. Micro-adjusting is also discussed in Chapter 2, in the context of individualized teaching.

Flexible grouping   We advocate an approach to mathematics teaching that includes both whole-class teaching and children working in small groups. Children are assigned to small groups in ways that take account of children's needs and the nature of the small group activity. Group membership might change frequently. Some activities are best suited to children working in pairs (for example, the games of Four Kings and Snap) and others are best suited to larger group sizes (for example, Bingo and Rabbit House tasks). For some activities it is appropriate to allocate children of similar levels of knowledge and learning needs to the same group. On other occasions it may be useful for a more advanced child to work with a child who is not as advanced in a particular aspect of number. To do so is often advantageous to both children. Through explanation and demonstration of strategies, more advanced children are likely to progress their knowledge of particular aspects of number.

Teaching at the cutting edge   Teaching is most effective when it is targeted just beyond the limit of a child's current number knowledge. In each lesson, learning activities such as tasks and games should be designed so that they are, on the one hand, challenging to more advanced thinkers and, on the other hand, not frustratingly difficult for less advanced thinkers. Teaching at the cutting edge and the related notion of zone of proximal development are discussed in Chapter 2, in the context of individualized teaching.

Children in a Problem-Centered Class

Develop their own methods   Children develop their own non-routine methods for solving problems as opposed to following a teacher-given set of procedures (Perry and Conroy, 1994). As stated above in the section on classroom discussion, children discuss and debate their solution strategies.

Reflect on their thinking   In a problem-centered classroom, children naturally think about their solutions and about the strategies used by their peers. When problems are posed, either to individuals or to the whole class, teachers allow adequate time for this reflection to occur and refrain from unwarranted attempts to provide help. On some occasions, rephrasing of a problem may be useful.

Expect to be challenged   There is an expectation on the part of children that learning mathematics involves engaging in challenging tasks. This is one of the norms that typically is established in a problem-centered classroom.

Climate of inquiry   In general, the aim of the above points is to foster an environment of inquiry and invention in mathematics. We contend that a classroom with a climate of discussion and questioning, where children's strategies and comments are valued by the teacher, will promote sense-making and intellectual autonomy, and foster positive attitudes in mathematics. Bringing about changes to the accepted ways (norms) of teaching and learning mathematics takes time and much effort. Nevertheless, establishing a climate of inquiry that incorporates the teaching and learning principles outlined above, is occurring in many classrooms.

Whole-Class Teaching

In this section we describe three different components (sessions) that can be incorporated into a typical lesson. These are short sessions, problem-solving sessions and specific activity sessions. As well, we provide an outline of a typical lesson and an example of a lesson using the outline. Also, the reader is referred to the authors' companion book, *Teaching Number in the Classroom with 4–8 year-olds* (Wright et al., 2006), which contains a detailed discussion of approaches to lesson organization and class management, and provides detailed sets of instructional activities for a range of early number topics.

Short sessions

Short sessions contain one or more brief activities. As well as being incorporated into a lesson, a short session might occur between other lessons or activities. Examples of activities that are suitable for short sessions are:

▶   saying forward number word sequences (beginning from a range of starting points)
▶   counting in sequences of multiples such as by 2s, 5s, and 10s
▶   saying backward number word sequences (beginning from a range of starting points)
▶   identifying numerals
▶   ascribing number to spatial patterns which are flashed (for example, dice patterns, doubles, five plus, ten plus, ten frame)
▶   building finger patterns (especially doubles and five-plus patterns).

Problem-Solving Sessions

The tasks used in problem-solving sessions are problems in the sense that the children do not have a ready-made solution or response. A problem is posed and children are given time to attempt a solution. This might involve working singly or in small groups. This is followed by a period of teacher-led discussion during which children describe, justify and compare their solutions. Problem-solving sessions have the following attributes:

▶   The problems can be solved at a range of levels and thus can engage children at different levels.
▶   The teacher guides the discussion.
▶   All children's strategies are valued.
▶   Class discussion is taken as the normal way of proceeding.
▶   The teacher does not comment on the correctness of a response but rather asks children to explain and justify their solutions.
▶   Incorrect responses are regarded as valuable opportunities to learn about children's thinking.
▶   The session ends with a short period during which children are encouraged to ask questions, clarify ideas and raise any related issues. During this period the teacher typically summarizes significant strategies and indicates where this session may lead in subsequent lessons.

In our view these sessions provide important opportunities for children to learn mathematics. These sessions typically involve using either one major problem only, or several smaller problems. Problems used in these sessions might involve addition, subtraction, tens and ones, multiplication, or division.

Specific Activity Sessions

During these sessions, children are given a specific activity, for example a game or task. The children might work singly or in small groups. These sessions might be incorporated into a lesson or occur separately. When incorporated into a lesson, the game or task typically complements the ideas being developed in the problems used in the lesson. When used separately from a lesson the teacher's purpose might be to have a particular child or small group engaging in a task or game in order to focus on some specific aspect of number. This is done to help overcome specific difficulties or to promote more advanced thinking. If another person (parent helper, support teacher or another child) is available to assist with the activity, the teacher should ensure that the helper has knowledge of the activity and its purpose. Where possible, the teacher should observe the work of helpers and discuss their work with them.

Outline of a Typical Lesson

Sessions of the kinds described above can be incorporated into a lesson. Following is an outline of a typical of lesson.

Step 1: Warm-Ups

Step 1 is based on the idea of a short session (see above). Warm-ups are provided through a series of short, focussed activities led by the teacher. Examples of these activities are identifying numerals, saying forward and backward number word sequences, identifying spatial patterns, and finger pattern activities.

Step 2: Posing the Problem

Step 2 is based on the idea of a problem-solving session (see above). A problem is posed to the whole class and discussed to clarify its meaning and to ensure all children are engaged. Similar problems might also be posed and discussed.

Step 3: Working in Small Groups or Individually

Step 3 is based on the idea of a specific activity session (see above). Activities for pairs or other small groups of children are given. These might include specific tasks similar to those given in the whole-class problem-solving time, games which supplement the problems given in the whole-class problem-solving time, or games and other activities designed for groups of children with a specific need. Some children might work individually rather than in a group.

Step 4: Plenary – Teacher-led, Whole-Class Discussion

During Step 4 the teacher assists children to think through the important elements of both the whole-class and small-group activities. The teacher might direct a child to describe an interesting or effective strategy. Alternatively, the teacher might discuss and connect several ideas relevant to the activities. Step 4 provides an opportunity for children to ask questions and talk about the game or activity that they have been involved in.

An Example of a Lesson

This is a second-grade lesson based on the lesson outline given above.

Step 1: Warm-Ups (about 6 minutes)

The teacher leads the children through the following activities.

Numeral identification task

Large cards, each of which has a numeral in the range 1 to 32, are held up one at a time.

Individual children are selected to name the numerals.

The teacher then displays some of the cards containing the 'teen' numerals and has the whole class state the name of each numeral.

Number word sequence tasks

The children say the number words forwards and backwards from various starting points.

Teacher: *Let's count from eight. Now let's count from forty-five.*

The teacher notes that there is some uncertainty moving from the fifties to the sixties and asks the class to say the number words again from fifty-eight to sixty-two.

Teacher: *Let's count backwards from sixteen.*

Many of the children have difficulty with saying the number words backwards through the teens.

The teacher then uses a numeral track to display each numeral one at a time and this assists the children in saying the number words from sixteen down.

The children are asked to state the number after and before a given number.

Teacher: *Who can tell me the number after —? (She holds up a numeral.)*

Spatial pattern tasks

Large cards (see Figure 3.4) looking like dominoes containing the 'five plus' patterns are quickly flashed.

Teacher: *How many dots are there? Show your answer by holding up the right number of fingers.*

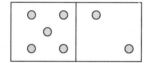 

Figure 3.4 Five Plus cards

After several cards are flashed.

Teacher: *Use your fingers to make a five. – Now make a seven. Make a five, now make a nine.*

Step 2: Posing the Problem (15 minutes)

Each child is given a small pack of digit cards (0, 1, ... 9). Each card contains one digit on one side and an adjacent digit on the other (for example, '4' on one side and '5' on the other). The children use the digit cards to make 2-digit numbers and thus display their answer (for example, using a '1' and a '5' to display '15'). Having different digits on each side makes it possible to make numbers such as 11, 22, 33, and so on.

The teacher places five cardboard bananas on the overhead projector (see Figure 3.5).

These are slid underneath a cardboard rectangle on the projector, which the teacher says is a banana box.

Teacher: *How many bananas are in the box?*

Three more bananas are placed next to the box (see Figure 3.5).

Teacher: *How many bananas are there altogether?*

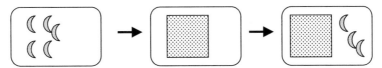

Figure 3.5 Going Bananas task

The children work out their answers and display the appropriate digit card. Some children display '8' quickly and others take more time. Some use fingers to obtain a solution. Quite a few children are looking at their neighbour's cards in order to check their answers. The teacher asks several children how they arrived at their answers and asks if anyone has a different way of working it out. Several children volunteer to tell the class about the strategy they used.

Notes   The teacher observes that there were five different strategies used by the children:

1. Many of the children counted the hidden bananas in the box: *One, two, three, four, five,* and then looked at the three visible bananas and said: *six, seven, eight. There are eight.*
2. Several children used a counting-on strategy: *There are five in the box.* Then looking at the other three bananas, saying, *six, seven, eight. There are eight bananas.*
3. One child said that in her mind she moved one banana out of the box, making four outside the box and four inside. She knew that four and four are eight.
4. Two children said they made the five on their fingers on one hand and put up three fingers on the other hand and they knew this made eight. The five plus combinations in the warm-up session may have promoted this strategy.
5. One child said he knew that five and three are eight because his sister told him. (Later questioning indicated to the teacher that this child had automated knowledge of many of the addition pairs.)

Several children needed to see the bananas in the box and had difficulty solving the task.

The teacher was encouraging in her comments to each of the children who described their strategies. She could see that most children were counting the hidden bananas from one and that counting-on would now be an appropriate strategy for the children to work towards. She therefore asked one of those children who used a counting-on strategy to explain once again how she worked out her answer. The two children who used the five plus finger strategy were invited by the teacher to come to the front of the class and demonstrate how they solved the problem. As these two children demonstrated their strategy, the remainder of the class followed, using the five plus strategy as well.

This discussion of strategies described above took about 8 minutes.

The teacher then proceeded to ask questions such as:

How many more will we need so that we have ten bananas in the box?

How many bananas would there be if there were two boxes, each having ten bananas? (This was not discussed in detail. The teacher planned to allocate these tasks in subsequent lessons.)

Step 3: Working in Small Groups or Individually (about 15 minutes)

The 'Going Bananas' game was demonstrated on the overhead projector. The instructions were as follows:

The class is organized into pairs.

Each pair has a box, two dice (one red, one green), a stack of banana tiles (made from 1 cm × 1 cm cardboard squares, each with a picture of a banana)

The green die has the numerals 4, 5, 6, 7, 8, 9 and the red die has the numerals 1, 2, 2, 3, 3 and 4 (alternatively red and green spinners could be used).

Each child has a banana tally sheet (see Figure 3.6).

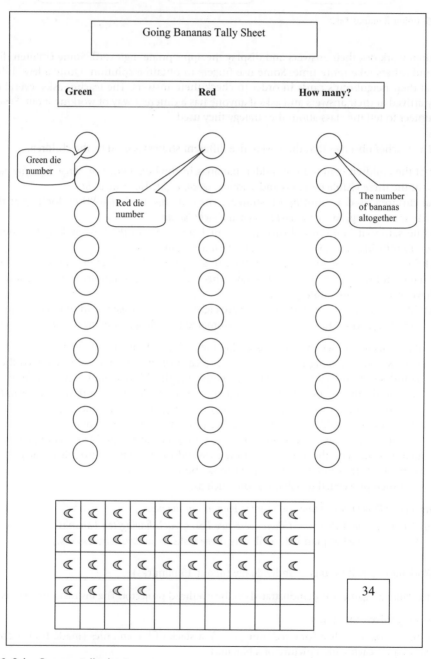

Figure 3.6 Going Bananas tally sheet

The first child throws the green die, takes this number of bananas and places them in the box. This number is written in the circle on the banana tally sheet in the green column. This same child throws the red die and places this number of bananas next to the box. This number is written in the circle in the red column on the tally sheet. (See Figure 3.6.) The child then works out the number of bananas altogether and writes this number on the tally sheet. The second child checks the result by counting the bananas if necessary. If correct, the first child places the bananas in the rows at the bottom of the tally sheet (see Figure 3.6). The second child then throws the dice and follows the same procedure and the first child does the checking. The game continues until the teacher calls a halt. The children write down the number of bananas they have altogether in the large square at the bottom of the tally sheet.

This game is designed to mirror the hidden addition tasks given to the whole class. It is useful if all of the children are involved in playing the game at the same time. How the children are grouped can vary. Children at similar levels of knowledge (for example, all counting-on) could play together. However, it might be productive to group together children at different levels.

Step 4: Plenary – Teacher-Led Discussion

The teacher assembles the children and orchestrates a discussion of the game. The children display their tally sheets. The teacher asks if anyone could see any fast ways of counting the bananas. She then holds up a bag and tells the children there are ten bananas in the bag. She then holds up three bananas and puts them in the bag and asks how many bananas are now in the bag. The teacher concludes the session by asking if anyone had learnt any different ways of working out how many bananas there are. Two children put up their hands. One says that they now use Carmel's way of doing it (which is to count on from the bananas in the box). The other says she does not like bananas and could we use apples next time!

HOW WILL I KNOW WHEN THEY GET THERE? CONTINUOUS ASSESSMENT

The LFIN (Chapter 1) enables teachers to continually assess their students' learning. When children describe their solutions, teachers should carefully note their responses. These responses allow teachers to continually micro-adjust their questions and activities during a lesson. The LFIN enables the teacher to assess at any time whether a child has attained a particular stage or level. As teachers become familiar with the framework: (a) continual assessment becomes part of their routine teaching practice; (b) they become skilled observers of children's strategies; and (c) they learn to pose tasks and questions which are likely to result in significant progress in children's learning of number.

In many of the schools we work with, teachers observe and document their children's progress in learning, in two complementary ways. First, they compare children's results on an initial assessment, as outlined earlier in this chapter, and a final assessment of similar format, conducted around the end of the school year. Second, they use a process of continual assessment as described in the previous paragraph.

SUMMARY

This chapter focussed first on the two instructional goals of sense-making and intellectual autonomy. Then followed a detailed description of the Teaching and Learning Cycle in four parts:

▶   *Where are they now?* This part relates to establishing a baseline from which teachers can accurately plan productive learning activities for their class as a whole and for individual children.
▶   *Where do I want them to be?* This part stresses the importance of teachers having a thorough knowledge of the mathematics they are teaching and a clear understanding of how children learn early number.

▶   *How will they get there?* This part highlights important principles for planning and attributes of problem-centered whole-class teaching. Several formats for whole-class teaching are described and an integrated lesson structure and example are provided.

▶   *How will I know when they get there?* This part refers to the importance of continual assessment and returns us to the beginning of the Teaching and Learning Cycle.

The LFIN is important for each aspect of the Teaching and Learning Cycle. It guides assessing, planning of lessons and lesson sequences as well as teaching.

4
General Introduction to Chapters 5 to 9

Following this introductory chapter there are five chapters, each of which focusses on teaching children whose numerical knowledge is assumed to be at a particular stage and particular levels. The focus of each chapter is determined primarily by a Stage of Early Arithmetical Learning. Thus Chapter 5 focusses on teaching children at the Emergent Stage (Stage 0), Chapter 6 focusses on teaching children at the Perceptual Stage (Stage 1), Chapter 7 focusses on teaching children at the Figurative Stage (Stage 2), Chapter 8 focusses on teaching children at the Counting-On and the Counting-Down-To Stages (Stages 3 and 4) and, finally, Chapter 9 focusses on teaching children at the Facile Stage (Stage 5). Thus from the perspective of instruction, Stages 3 and 4 are combined. This is because, for the child who has reached the Counting-On Stage (Stage 3), it is not considered crucial to focus teaching on the development of counting-down-to, that is, the strategy characteristic of Stage 4. Rather, when children have developed robust counting-on and counting-back strategies, teaching should focus on advancing the child to the facile stage (that is, Stage 5).

As well as assuming for each chapter a specific Stage of Early Arithmetical Learning, it is also assumed for each chapter that the child is at a specific level on other key aspects of early arithmetical knowledge. Thus in Chapter 5 for example, it is assumed that the child is at Level 1 in terms of the model of FNWSs, and Level 0 in terms of the models of BNWSs, Numeral Identification and Tens and Ones Knowledge. It should be kept in mind that children at a particular stage can vary in terms of their levels on the other models. Table 4.1 shows the levels assumed on each of six models of early arithmetical knowledge. Each of these assumed levels is explained in detail in the relevant chapter.

Table 4.1 Stages and levels assumed for a typical child in Chapters 5–9

Model	Chapter				
	5	6	7	8	9
Stage of Early Arithmetical Learning (SEAL)	0	1	2	3	5
Level of Forward Number Word Sequences (FNWSs)	1	3	4	5	5
Level of Backward Number Word Sequences (BNWSs)	0	l	3	4	5
Level of Numeral Identification	0	1	2	3	4
Level of Tens and Ones Knowledge	0	0	0	1	2
Level of Early Multiplication and Division Knowledge	—	—	—	1	3

RATIONALE FOR ASSUMING SPECIFIC LEVELS

The purpose of this section of the book is to provide very specific guidance for teaching to advance children's early numerical strategies. In order to do this it is necessary that the instructional approaches set out here are closely attuned to children's current levels of knowledge. In order to achieve this it is convenient to assume a typical profile of early number knowledge for each chapter. Of course, knowing that a child is at a given stage in terms of SEAL does not necessarily determine the child's levels in terms of the other models of early number knowledge. Nevertheless it is feasible to assume levels on the other models that are typical of a child at a given stage. As has been emphasized in earlier chapters, there is no substitute for knowing in as much detail as possible, the child's current levels of knowledge.

CHAPTER FORMAT AND OVERVIEW

Each of Chapters 5–9 is set out according to the following format: an introductory section that provides a description of the typical child whose levels are those assumed for that chapter; descriptions of six key topics each of which is considered to be an important focus of instruction for a child at the assumed levels; and examples of whole-class lessons designed for children at the assumed levels. The whole-class lessons are intended to exemplify approaches to classroom teaching considered appropriate for a class of children who are considered to be in general terms, at the levels specified for that chapter. Each lesson is similar in focus and content to one or more of the key topics for that chapter. For each lesson the key topics related in this way are listed at the top of the lesson under the heading 'Links to Key Topics'.

DESCRIPTIONS OF THE TYPICAL CHILD FOR EACH CHAPTER

The purpose of this section in each chapter is to provide additional information about the number knowledge of the typical child, that is, additional to the information implied by the assumed stage and levels. This additional information is organized in terms of various aspects of the Learning Framework in Number. In Chapter 5 for example, there is additional information about the following aspects: early arithmetical strategies, FNWSs, BNWSs, numeral identification, spatial patterns, finger patterns and temporal sequences. In Chapter 9 for example, there is additional information about early arithmetical strategies, grouping by fives and tens, FNWSs, BNWSs, numeral identification, tens and ones, and early multiplication and division. This additional information should enable the reader to gain a detailed picture of the children for whom the key teaching topics in the chapter are intended.

KEY TOPICS – FOCUS AND CATEGORY

In each chapter, each key topic addresses an important area of instructional focus for children whose levels of early numerical knowledge are those assumed for the chapter. For Chapters 5–9 the key topics fall into one of the following three categories: (a) Number words and numerals, (b) Counting, and (c) Grouping. These categories correspond with the three broad phases of early numerical knowledge which are described in detail in Chapter 1. Thus the key topics in the category of Number Words and Numerals are intended mainly to address instruction to advance children's knowledge of number words and numerals. In similar vein, each of the key topics categorized as Counting/Grouping is intended mainly to address instruction to advance children's knowledge in the corresponding category.

KEY TOPICS – FORMAT

Each key topic is summarily described by its title and purpose which appear at the top of the section for that key topic. Table 4.2. sets out the titles and purposes of the key topics for each of the five chapters. Also indicated in Table 4.2 are the code and category for each key topic, and the number of teaching procedures in the key topic. The description of each key topic also includes the following: (a) Links to LFIN; (b) Teaching procedures; (c) Vocabulary; (d) Materials; and (e) Acknowledgments. The Links to LFIN has the purpose of locating the key topic in the Learning Framework in Number which is described in Chapter 1. The section headed 'Teaching Procedures' constitutes the main part of each key topic and is explained in detail in the next paragraph. The section headed 'Vocabulary' has the purpose of highlighting key words and terms used in the teaching procedures of the key topic and the section headed 'Materials' lists the required instructional materials. Many of the teaching procedures and instructional materials have been adapted from other sources and this is indicated for each key topic under the heading 'Acknowledgments'.

Table 4.2 Titles and purposes of key topics for Chapters 5–9

W = Number words and numerals; C = Counting; G = Grouping; No. = Number of key topics

Code and category	Chapter and title	Focus	No.
	Emergent		
5.1-W	Number Word Sequences from 1 to 20	Knowledge of forward number word sequences in the range 1 to 20 and backward number word sequences in the range 1 to 10	7
5.2-W	Numerals from 1 to 10	Knowledge of numerals and numeral sequences in the range 1 to 10	6
5.3-C	Counting Visible Items	Perceptual counting strategies	5
5.4-G	Spatial Patterns	Initial facility to ascribe number to spatial patterns and random arrays	3
5.5-G	Finger Patterns	Initial facility with making finger patterns	7
5.6-C	Temporal Patterns and Temporal Sequences	Facility with copying and counting temporal patterns and temporal sequences	4 = 32
	Perceptual		
6.1-W	Number Word Sequences from 1 to 30	Knowledge of number word sequences in the range 1 to 30	7
6.2-W	Numerals from 1 to 20	Knowledge of numerals in the range 1 to 20	6
6.3-C	Figurative Counting	Figurative counting strategies	4
6.4-G	Spatial Patterns	Facility to ascribe number to regular spatial patterns	5
6.5-G	Finger Patterns	Facility with finger patterns for numbers in the range 1 to 10	4
6.6-G	Equal Groups and Sharing	Initial ideas of equal groups and equal sharing	6 = 32

▶

Table 4.2 Continued

W = Number words and numerals; C = Counting; G = Grouping; No. = Number of key topics

Code and category	Chapter and title	Focus	No.
	Figurative		
7.1-W	Number Word Sequences from 1 to 100	Knowledge of number word sequences in the range 1 to 100	7
7.2-W	Numerals from 1 to 100	Knowledge of numerals in the range 1 to 100	10
7.3-C	Counting-On and Counting-Back	Counting strategies involving counting-on and counting-back	7
7.4-G	Combining and Partitioning Involving Five and Ten	Facility with using five and ten to combine and partition numbers in the range 1 to 10	6
7.5-G	Partitioning and Combining Numbers in the Range 1 to 10	Facility with partitioning and combining numbers in the range 1 to 10	7
7.6-G	Early Multiplication and Division	Early multiplicative and divisional strategies	6 = 43
	Counting-On		
8.1-W	Number Word Sequences by 2s, 10s, 5s, 3s and 4s	Facility with forward and backward number word sequences by 2s, 10s, 5s, 3s and 4s in the range 1 to 100	8
8.2-W	Numerals from 1 to 1,000	Knowledge of numerals in the range 1 to 1,000	7
8.3-C	Incrementing by Tens and Ones	The facility to increment and decrement numbers by tens and ones, in the range 1 to 100	4
8.4-G	Adding and Subtracting to and from Decade Numbers	The facility to add numbers in the range 1 to 9, to and from decade numbers, and to subtract numbers in the range 1 to 9, to and from decade numbers	6
8.5-G	Addition and Subtraction to 20, using 5 and 10	Facility with addition and subtraction in the range 1 to 20, using grouping by 5 and 10	9
8.6-G	Developing Multiplication and Division	Early multiplicative and divisional strategies	7 = 41
	Part-whole		
9.1-W	Counting by 10s and 100s	Facility with counting forwards and backwards by tens off the decade, and counting by 100s on and off the 100, and on and off the decade	6
9.2-C	2-Digit Addition and Subtraction through Counting	Counting-based strategies for 2-digit addition and subtraction	7
9.3-G	Non-Canonical Forms of 2-Digit and 3-Digit Numbers	Facility to associate non-canonical forms of 2-digit and 3-digit numbers with their canonical forms	5
9.4-G	2-Digit Addition and Subtraction through Collections	Collections-based strategies for 2-digit addition and subtraction	6
9.5-G	Higher Decade Addition and Subtraction	Strategies for adding numbers in the range 2 to 9, to 2-digit numbers, and subtracting numbers in the range 2 to 9, from 2-digit numbers	4
9.6-G	Advanced Multiplication and Division	Advanced multiplicative and divisional strategies	6 = 34
			182

TEACHING PROCEDURES

The teaching procedures have the purpose of providing exemplars of teaching which are closely attuned to the stage or a level of early numerical knowledge assumed for that chapter. In each teaching procedure, exemplary teacher dialogue and action are listed, with teacher dialogue indicated by italics. The teaching procedures are intended to be illustrative and are not intended necessarily to be followed verbatim. The authors' intention is that the procedures should be used by teachers in situations involving individualized teaching. As well, the procedures can be adapted for use in situations involving group or class teaching.

NOTES ON PURPOSE, TEACHING AND CHILDREN'S RESPONSES

At the end of each teaching procedure a paragraph headed 'Purpose, Teaching and Children's Responses' is provided. Each paragraph sets out a list of notes that are important to fully understand and use the corresponding teaching procedure. Notes on purpose and teaching help the teacher to determine an appropriate focus and points of emphases for the procedure, and descriptions of likely child responses can prepare the teacher for possible adaptations and extensions as they proceed.

SAMPLE LESSONS

Each chapter culminates with examples illustrating how one or more of the key topics can be built into lessons. The format of the lessons is similar and can be summed up under Title; Purpose; Links to Key Topics; Materials; Introductory Activities; Main Focus; Group/Individual Tasks; Conclusion/Summary. The layout is therefore applicable to lessons in general and matches the CMIT project in Australia and the style of lessons advocated by the National Numeracy Strategy in England and Wales.

IN CONCLUSION

The format of each of the Chapters 5–9, Teaching the Emergent to Facile child is summarized in Table 4.3.

Table 4.3 Summary formats of Chapters 5–9

The typical profile of a child at Stage X
Additional information to explain the number knowledge of the child to present the reader with a detailed picture of the children for whom the topics are intended

The six key topics
The topics which are considered to be the important focus of teaching at the assumed levels for the chapter

Description of key topics
Title
Purpose
Links to LFIN: locating the key topic within the LFIN
Illustrative teaching procedures closely attuned to the stage and levels addressed   ▶

Table 4.3 Continued

(a) suggested phraseology, and/or action for the teacher
(b) further explanation of the objectives to provide a focus and identify points for emphasis
(c) examples of probable pupil responses, together with possible errors, which the teacher may anticipate and therefore have adaptions or extension activities ready
(d) vocabulary, highlighting main words and terms
(e) instructional materials
(f) acknowledgment of sources

Examples of whole-class lessons designed for children at the assumed stage and levels
Title
Purpose
Links
Materials
Introductory activities
Main Focus
Activities for groups and individuals
Conclusion/summary

5
Teaching the Emergent Child

This chapter focusses on teaching children at the Emergent Stage on the SEAL (Stages of Early Arithmetical Learning – see Chapter 1). First, a detailed description of the knowledge and strategies typical of this stage is provided. This is followed by detailed descriptions of six key topics which can provide a basis for teaching the child at the Emergent Stage. The six key topics consist of a total of 32 teaching procedures.

THE TYPICAL EMERGENT CHILD

This section provides an overview of a typical child at the Emergent Stage, that is, at Stage 0 of the Stages of Early Arithmetical Learning. Table 5.1 sets out the stage and indicative levels for the Emergent child on the models pertaining to FNWSs, BNWSs, Numeral Identification and Tens and Ones. The overview discusses the first four aspects of early number knowledge listed in Table 5.1, and also discusses facility with finger patterns and knowledge associated with spatial and temporal patterns.

Table 5.1 Stage and levels of a typical emergent child

Model	Stage/level
Stage of Early Arithmetical Learning (SEAL)	0
Level of Forward Number Word Sequences (FNWSs)	1
Level of Backward Number Word Sequences (BNWSs)	0
Level of Numeral Identification	0
Level of Tens and Ones Knowledge	0

Stage of Early Arithmetical Learning

The child at the Emergent Stage is not able to count a collection of counters, for example 13 or 18 counters. The child might not know the forward number word sequence from one to ten or beyond ten. Alternatively, the child might know the number words but not be able to correctly coordinate a number word with each counter. This might involve omitting one or more counters or perhaps pointing to one or more counters twice during counting. More frequently the child simply lacks the ability to coordinate each spoken number word with each item to be counted. In many cases the child's pointing actions seem to outpace their production of the number words in turn. The child might be able to count the counters in smaller collections, collections up to six or 10 for example. Some emergent children seem to interpret questions such as, 'how many counters are there?', literally as an instruction to say the number word sequence from one. It is as if the literal meaning for the child is to say the words 'one, two, three, and so on'. while pointing at the collection of items in question. This might involve pointing generally in the direction of the collection, rather than pointing at each item in turn.

FNWSs, BNWSs and Numeral Identification

The emergent child might well be able to say the forward number word sequence from one to beyond ten, nevertheless they probably will not be able to say immediately the number word after a given number word in the range one to ten. As well, they probably will not be able to use a dropping back strategy to say the number word after a given number word, for each of the numbers in the range one to 10. The emergent child typically has difficulty with saying backward number word sequences. For example, the child might not be able to say the words from ten to one, and might even have difficulty in saying the words from three to one. The child typically cannot say the number word before a given number word, even permitting use of the dropping back strategy. The emergent child can typically name some but not all of the numerals in the range 1 to 10. They might be able to recognize most of the numerals in the range 1 to 5, but might be unable to name numerals beyond five, or might confuse numerals such as 6 and 8 or 6 and 9. In the case of handwritten numerals, they might confuse numerals such as 2 and 6.

Spatial Patterns, Finger Patterns and Temporal Sequences

The emergent child might be able to recognize some but not all of the regular spatial patterns (for example, the domino patterns) in the range 2 to 6. In many cases they will try to count the dots in a spatial pattern rather than immediately assign a number name to the pattern. The emergent child might be able to make finger patterns corresponding to the numbers from one to five. This typically will involve looking at the fingers and raising them slowly and sequentially, for example the child might slowly raise three fingers sequentially in response to a request to show three fingers. The emergent child might be able to copy or count temporal sequences of two or three but no larger (for example, on hearing a sequence of three sounds the child could tap a similar sequence).

The Way Forward

It is important to realize that the emergent child has a significant basis of early number knowledge on which further number knowledge can be built. An appropriate instructional programme will focus simultaneously on: (a) strengthening facility with forward number world sequences in the range 1 to 20; (b) strengthening facility with backward number words sequences in the range 1 to 10; (c) extending various aspects of knowledge of numerals – numeral recognition and identification, and numeral sequences; (d) ascribing numerosity to spatial patterns, temporal patterns and temporal sequences; and (e) making finger patterns for numbers in the range one to five.

The next section contains six key topics and 32 teaching procedures which form the basis of an appropriate instructional programme for the emergent child.

Key Topic 5.1: Number Word Sequences from 1 to 20

Purpose: To develop knowledge of forward number word sequences in the range 1 to 20 and backward number word sequences in the range 1 to 10.

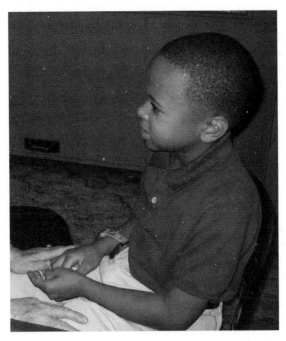

Saying the number word sequence

LINKS TO LFIN

Main links: FNWS, Levels 1–4; BNWS, Levels 1–3.
Other links: SEAL, Stage 1.

TEACHING PROCEDURES

5.1.1: Copying and Saying Short FNWSs

▶ *I'm going to count from one to three, and I want you to say it after me. Ready; one, two, three. Again; one, two, three.*

▶ *Now, I'm going to count from four to six, and I want you to say it after me. Ready; four, five, six. Again; four five, six.*

▶ Similarly for the number words from seven to ten.

▶ *I'm going to count from one to five, and I want you to say it after me. Ready; one, two, … five. Again; one, two, … five.*

▶ Similarly for the number words from six to ten.

▶ *This time, you count from one to five by yourself.*

▶ Similarly, six to ten, one to ten.

▶ Extend to FNWSs in the range one to 20.

Purpose, Teaching and Children's Responses

▶ Some children will know the number word sequence to five or six only.
▶ This teaching procedure can serve to extend the range in which the number word sequence is known.
▶ Ensure that you give children enough time to solve tasks and reflect on their responses.
▶ Many children will initially have difficulty saying FNWSs that do not start from one (for example, starting from four, starting from six).
▶ Children might have difficulty in pronouncing 'thirteen', 'fourteen', ... 'nineteen' and might confuse these with decade names ('thirty' and so on).

5.1.2: Copying and Saying Short BNWSs

▶ *I'm going to count backwards from three, and I want you to say it after me. Ready; three, two, one.*
▶ *Now I'm going to count backwards from six, and I want you to say it after me. Ready, six, five, four.*
▶ *Now I'm going to count backwards from ten, and I want you to say it after me. Ready, ten, nine, eight.*
▶ Similarly four to one, and eight to five, ten to seven.
▶ *This time, you count from five back to one by yourself.*
▶ Similarly, eight to three, ten to five, ten to one.

Purpose, Teaching and Children's Responses

▶ BNWSs are usually significantly more difficult for children than FNWSs.
▶ Children at this level will frequently use a dropping back strategy in order to produce a required number word in a BNWS. Thus the child says an FNWS subvocally to determine the next number word in a BNWS, for example, the child says subvocally 'one, two, three, four, five' to determine what comes before 'five'.
▶ Ensure children have enough thinking time and have an opportunity to reflect on their solutions.

5.1.3: Saying Alternate Number Words Forwards and Backwards

▶ *Let's take turns to say the numbers. I will say one, then you say two, then I will say three, then you say four, and we will keep going like that. Ready, one, (two), three, (four), ...*
▶ *This time you start with one. Ready, (one), two, (three), four, ...*
▶ *Let's try that going backwards. Let's go backwards from six. I'll start off. Ready, six, (five), four, ...*
▶ *This time we will go backwards and you can start from eight. Ready, (eight), seven, (six), —.*
▶ Similarly with FNWSs to twenty and BNWSs to ten.

Purpose, Teaching and Children's Responses

▶ In this activity the child reflects on the last number word said by the teacher in order to figure out what word to say next. The process of reflection is somewhat different from that in 5.1.1 and 5.1.2 above for example. Reflection on spoken number words in these various ways can strengthen the child's knowledge of FNWSs and BNWSs.
▶ If the child has continuing difficulty with saying BNWSs, have the child repeat short sequences after the teacher.

5.1.4: Saying the Next Number Word Forwards

▶ *I'm going to count, and I want you to say the next number after I stop. Ready, one, two, three, four, —. Five, six, seven, eight, —.*

▶ *I'm going to count, and I want you to say the next number after I stop. Ready, one, two, —. Five, six, —. Eight, nine, —.*

▶ Extend to number words in the range 11 to 20.

Purpose, Teaching and Children's Responses

▶ These activities and those below (see 5.1.5, 5.1.6 and 5.1.7) constitute an important basis for advanced counting strategies (counting-on and counting-down) to be developed later.

▶ Children are likely to encounter more difficulty when fewer number words are said by the teacher. Thus 'seven, eight, —', is likely to be more difficult than 'five, six, seven, eight, —'.

▶ For this teaching procedure and the next (5.1.5) it is important that the child listens carefully to the teacher, and thinks carefully before answering.

▶ Choose the difficulty levels of the tasks carefully. If too many of the tasks are too difficult the child might resort to guessing without thinking very much.

5.1.5: Saying the Next Number Word Backwards

▶ *I'm going to count backwards, and I want you to say the next number backwards, after I stop. Ready, four, three, two, —. Seven, six, five, —.*

▶ *I'm going to count backwards, and I want you to say the next number backwards, after I stop. Ready, five, four, —. Eight, seven, —. Nine, eight, —.*

Purpose, Teaching and Children's Responses

▶ As in the case of next number word forwards, children are likely to encounter more difficulty when fewer number words are said by the teacher Thus 'eight, seven, —,' is likely to be more difficult than 'ten, nine, eight, seven, —'.

5.1.6: Saying the Number Word After

▶ *I'm going to say a number and I want you to say the number just after the one I say. Ready, seven! Four! Eight!*

▶ Extend to number words in the range 11 to 20.

Purpose, Teaching and Children's Responses

▶ Initially, children are likely to use a subvocal dropping back strategy, for example, the child says 'one, two, ... six, seven, eight' to figure out what comes after 'seven'.

▶ When this teaching procedure and the next (5.1.7) are used frequently together, it is not unusual for children to confuse number after and number before (for example, answering 'seven' for 'what comes before six?'). In such cases one should present tasks involving FNWSs only (see 5.1.1, 5.1.4 and 5.1.6 above) for one or more lessons as necessary.

▶ These procedures can be used in conjunction with procedures involving numeral sequences (Key Topic 5.2).

5.1.7: Saying the Number Word Before

▶   *I'm going to say a number and I want you to say the number before the one I say. Ready, four! Ten! Three!*

Purpose, Teaching and Children's Responses

▶   As in the case of number word after (see 5.1.6), the child might subvocally use a dropping back strategy.

VOCABULARY

One, two, three, ... twenty, count from, count to, number, forwards, backwards, next number after, next number backwards, number just after, number before, stop, forwards from, backwards from, start from.

MATERIALS

None.

ACKNOWLEDGMENT

These activities were adapted from the work of Robert Wright, and were further developed in the Mathematics Recovery project.

Key Topic 5.2: Numerals from 1 to 10

Purpose: To develop knowledge of numerals and numeral sequences in the range 1 to 10.

LINKS TO LFIN

Main links: Numeral Identification, Level 1.
Other links: FNWS and BNWS, Levels 1–3; SEAL, Stage 1.

TEACHING PROCEDURES

5.2.1: Numeral Sequences, Forwards

▶   Place out the numeral sequence from 1 to 3. *Here are some numbers. Watch me as I count from one to three.* Point to each numeral in turn, while counting from one to three. *One, two, three.*
▶   *Now you say the numbers with me from one to three.*
▶   *Now you say the numbers and point to each one like I did before.*
▶   Similarly, one to four, one to five, one to six, ... one to ten.

Purpose, Teaching and Children's Responses

▶ These activities are complementary to those in Key Topic 5.1.
▶ In most of the teaching procedures in this key topic (5.2) the child's emerging knowledge of FNWSs and BNWSs supports learning of numerals and their emerging knowledge of numerals supports learning of FNWSs and BNWSs. These are important interrelated aspects of early number learning (see Chapter 1).
▶ Children might initially have difficulty in coordinating each point with saying the corresponding number word, for example, they might say the words from one to five, while pointing at only two or three of the numerals. In such cases the teacher should model the activity slowly for the child.

5.2.2: Numeral Sequences, Forwards and Backwards

▶ Place out the numeral sequence from 1 to 3. *Watch me as I count forwards and backwards.* Point to each numeral in turn, while counting from one to three, and then from three to one. *One, two, three; three, two, one.*
▶ *Now you say the numbers with me as we count from one to three and from three to one.*
▶ *Now you say the numbers and point to each one like I did before.*
▶ Similarly, forwards and backwards: one to four, one to five, one to six, ... one to ten.

Purpose, Teaching and Children's Responses

▶ Children might have difficulty after saying the last number word forward and then attempting to say the first number word backward. In the case of one to four, they might say for example, 'one, two, three, four, three, two, one' or 'one, two, three, four, five, six, —'.
▶ As before, for some children, careful modeling by the teacher is necessary.

5.2.3: Sequencing Numerals

▶ Place out the cards from 1 to 3, randomly arranged. *Put these cards in order from one.* Direct the child to arrange cards from left to right in increasing order. *Now say the numbers as you point to them.*
▶ Similarly order the cards from 1 to 4, 1 to 5, 4 to 7, 6 to 9, 6 to 10, and so on.

Purpose, Teaching and Children's Responses

▶ Children are likely to encounter significantly more difficulty when there are more cards, for example, ordering 1–3, is much easier than ordering 1–5, which in turn is easier than ordering 1–10.
▶ Observe how children put the cards in order. Some children might spontaneously order the cards vertically (top to bottom or bottom to top) rather than from left to right. Some might have a preference for ordering right to left.
▶ The teacher should determine the range of numeral cards which is just beyond the range which the child can order without difficulty.
▶ It is important to provide sufficient time for the child to think hard – some children will require several seconds to identify or read some of the numerals. Thus this task involves the child identifying numerals and then thinking about the corresponding FNWS, so as to sequence the numerals.
▶ In the case of sequences that do not start from 1 (for example, sequencing 6–10), some children might initially appear to be confused because they expect to see the numeral 1 in the group of numerals to be sequenced.

5.2.4: Numeral Recognition

▶ Place out the cards from 1 to 3, randomly arranged. *Point to the number 2. Point to the number 3. Point to the number 1.*
▶ Similarly using the cards from 1 to 4, 1 to 5, 4 to 7, 6 to 10, and so on.

Purpose, Teaching and Children's Responses

▶ Some children can recognize the numerals from 1 to 5, but have significant difficulty with recognizing the numerals from 6 to 10.
▶ Some children might frequently confuse particular numerals (for example, confusing 6 and 8).

5.2.5: Numeral Identification

▶ Place out the cards from 1 to 3, randomly arranged. Point to the card for 2. *What number is this?* Point to the card for 3. *What number is this?*
▶ Similarly using the cards from 1 to 4, 1 to 5, 4 to 7, 6 to 10, and so on.
▶ Place the cards from 1 to 5 in a pile, face-down and randomly arranged. *I am going to turn over each card in turn. Tell me the number on the card. What number is this?*
▶ Similarly using the cards from 1 to 6, 1 to 8, 1 to 10.

Purpose, Teaching and Children's Responses

▶ Some children will say the FNWS from one, aloud or subvocally, in order to generate the name of a numeral.
▶ As for Teaching Procedure 5.2.4 (see above) some children might frequently confuse particular numerals. In such cases, visual discrimination tasks can be useful. For example, use a pile of cards where either '6' or '8' appears on each card. The child's task is to sort the cards into two piles (6s and 8s).

5.2.6: Numeral Tracks – 1 to 5 and 1 to 10

▶ Place out the numeral track from 1 to 5, with numerals uncovered. *Watch me as I count forwards and backwards.* Point to each numeral in turn, while counting from one to five, and then five to one. *Now you count forwards and backwards and point to each number in turn.*
▶ Close lids on the numeral track and repeat previous activity, uncovering the lids after saying each number.
▶ Place out the numeral track from 1 to 5, with all numerals covered. Uncover the numeral 3. *What number is that?* Leave the numeral 3 uncovered, and point to the lid covering the numeral 4. *What number is that?* Direct the child to uncover the numeral 4. *Were you correct?*
▶ Similarly for other numerals from 1 to 5.
▶ Repeat the previous activities using the numeral track from 1 to 10. Other numeral tracks in the range 1 to 10 may be used, for example, 1 to 6, 1 to 8, 6 to 10.

Purpose, Teaching and Children's Responses

▶ The activity of turning over the appropriate lid after naming a numeral seems to be enjoyable for many children. This is because of the self-checking aspect of uncovering the numeral in question.

▶ Children frequently use a strategy of counting forwards subvocally or aloud from the start of the numeral track, for example, if the numeral 7 is uncovered, and the child is asked to name the next numeral (that is, 8), the child might count the lids (that is, one, two, …) from the start of the track.

▶ After working with a particular numeral track children will come to know where each numeral is located on the track. Thus they associate a numeral name (that is, a number word) with each lid on the numeral track. We do not regard this as problematic, but teachers should be aware of this when it is happening. This kind of teacher observation and awareness is an important part of understanding children's strategies and the extent of their knowledge.

▶ Related to the previous point (that is, the child associates a numeral name with each lid on the numeral track), when this happens it is likely that the child has developed a reasonably robust mental image of the numeral sequence, and this mental image can be used in solving certain tasks.

VOCABULARY

Number, count forwards, count backwards, in order, point, numeral track, lid, cover, uncover

MATERIALS

Numeral sequences from 1 to 3, 1 to 4, 1 to 5, … 1 to 10.
Numeral cards for each number in the range 1 to 10.
Numeral track from 1 to 10.

ACKNOWLEDGMENTS

These activities were adapted from the work of Robert Wright, and were further developed in the Mathematics Recovery project. The numeral track was developed by Garry Bell, Southern Cross University.

Key Topic 5.3: Counting Visible Items
Purpose: To develop perceptual counting strategies.

LINKS TO LFIN

Main links: SEAL, Stage 1.
Other links: FNWS and BNWS, Levels 1–4.

TEACHING PROCEDURES

5.3.1: Counting Items in One Collection

▶ Place 8 red counters in a group. *How many counters are there?*
▶ Similarly, 12 counters, 15 counters, and so on.

Purpose, Teaching and Children's Responses

▶ Typically children will point at each counter in turn, or move each counter in turn, although some children do not spontaneously point at or move the counters.
▶ Initially children might encounter difficulty in coordinating a number word with each counter, for example, they might point at two counters when saying 'sev-en' or they might point at only one counter when saying 'five, six'.
▶ Children might make an error with the number word sequence, for example they might omit 'eleven' or 'thirteen'.
▶ Some children might not realize that, when counting, the last number word is the answer to 'How many'?
▶ Related to this, some children seem to think literally that the appropriate response to the question 'How many?' is to say the number words from one, while pointing at each counter in turn. In such cases it can be useful for the teacher to model for the child, the activity of ascribing number to small collections without counting from one, for example, display two counters and say 'two' (similarly 3, 4, and so on).

5.3.2: Establishing a Collection of Given Numerosity

▶ Place out a group of about 30 red counters. *Get me 6 counters from the group.*
▶ Similarly, 10 counters, 14 counters, and so on.

Purpose, Teaching and Children's Responses

▶ Some children may not fully realize that the task involves separating some of the counters from the initial group of counters. Having this realization is referred to as having a 'pull-out' strategy.
▶ Even if the child initially separates the counters into two groups, they might not realize that one of the groups constitutes the group of counters that has been requested.
▶ Errors of coordination (see 5.3.1 above) can occur on these tasks as well.
▶ As before, teacher modeling with small collections can be useful (*I am going to get 3 counters. Now I am going to get 5 counters. And so on.*).

5.3.3: Counting Items in a Row, Forwards and Backwards

▶ Place out a row of six dots. *Watch me as I count the dots, forwards and backwards.* Point to each dot in turn. *One, two, … six! Six, five, … one!*
▶ *Now you count the dots, forwards and backwards.*
▶ Similarly, 9 dots, 13 dots, and so on.

Purpose, Teaching and Children's Responses

▶ Some children will have difficulty initially in counting backwards and realizing that, for example, after completing the forward count, the last uttered number word is repeated to commence the backward count.
▶ Because the counters are in a row and because the task involves counting backwards as well as forwards, the tasks can highlight for the child that a number word is associated with each counter. Thus the task can serve to reinforce the idea of associating one number word only with each item when counting (that is, counting-by-ones).

5.3.4: Counting Items of Two Collections

▶ Place 6 red counters in a group. Place 4 green counters in a group. *Here are 6 red counters and 4 green counters. How many counters are there altogether?*

▶ Similarly, 8 red and 2 green, 12 red and 1 green, 5 red and 3 green, 14 red and 2 green, and so on.

Purpose, Teaching and Children's Responses

▶ It is not uncommon for children to have difficulty realizing that two collections can be alternatively regarded as one collection for the purposes of counting how many counters in all. Thus in the case of 6 red counters and 4 green counters, they will answer 'six, four' when asked how many altogether. In such cases the teacher can model the action of physically reorganizing the counters into one group and then counting from one, all of the counters in the newly formed group.

▶ The mental act of conceiving of two physically distinct groups (or otherwise distinct, for example, distinguished by color) as one group for the purposes of counting is one that has to be learned by some children at this level. For such children the tasks in this teaching procedure can be quite important.

5.3.5: Counting Items of Two Rows

▶ Place out a row of 10 red dots. Beside that place a row of 3 green dots. *Here are 10 red dots and 3 green dots. How many dots altogether?*

▶ Similarly, 8 red and 3 green, 12 red and 4 green, and so on.

Purpose, Teaching and Children's Responses

▶ In similar vein to points made above (see 5.3.1 to 5.3.4) children might incorrectly coordinate number words and dots or might not realize that the dots in the two rows can be alternatively regarded as one group of dots for the purposes of counting how many in all.

VOCABULARY

Counters, group, collection, how many, altogether, number, count, row, dots, forwards, backwards, red counters, green counters.

MATERIALS

30 red counters, 20 green counters, rows of 6, 10, 15 and 20 red dots, rows of 1, 2, 3, 4, 5 green dots.

ACKNOWLEDGMENT

Most of these activities were adapted from the work of Leslie Steffe and the work of Robert Wright.

Key Topic 5.4: Spatial Patterns

Purpose: To develop initial facility to ascribe number to spatial patterns and random arrays.

LINKS TO LFIN

Main links: Part C, Spatial Patterns.
Other links: FNWS and BNWS, Levels 1–3; SEAL, Stage 1.

TEACHING PROCEDURES

5.4.1: Ascribing Numerosity to Patterns and Random Arrays

▶ Display domino card for 1. *How many dots do you see?* Similarly for domino cards for 2, 3, and 4 in order.
▶ Display cards in random order. *How many dots do you see?*
▶ *Tell me how many dots you see. Ready!* Flash cards for 1, 2, 3, 4 in turn.
▶ *Let's do that again. Ready!* Flash cards for 1 to 4 in random order.
▶ Repeat the above sequence using random array cards for 1 to 4.
▶ Repeat the above sequence using domino cards for 1 to 6.
▶ Repeat the above sequence using pairs patterns for 1 to 6.

Purpose, Teaching and Children's Responses

▶ In the initial case when the card is displayed rather than flashed, many children will count the dots slowly or carefully from one. In this case children might make coordination errors, might omit to count one or more dots or might count one or more dots more than once.
▶ In the cases when the card is flashed, the child might immediately recognize the pattern in the sense of immediately recalling the number name that corresponds to the pattern. Alternatively, the child might take a few seconds to answer, in which case they might be visualizing the pattern and attempting to count the dots in their visualized image.
▶ An important goal is for children to develop strong conceptually based spatial patterns. The term 'conceptually based' is used to indicate that the child can visualize a pattern for 'four' for example, in the absence of any visibly available pattern (that is, a pattern for four is neither displayed nor flashed).
▶ These conceptually based patterns can form a basis for children emerging strategies for combining and partitioning small numbers (for example, numbers in the range 1 to 10).

5.4.2: Making Spatio-Motor Patterns to Match Spatial Patterns

▶ Display domino card for 2. *Make a pattern in the air to show the number of dots.*
▶ Display domino card for 3. *Make a pattern in the air again but this time see if you can do it without looking at the card.*
▶ Similarly for domino cards for 1 and 4.
▶ *This time I am going to flash the pattern. Look at the pattern and make a pattern in the air to show the number of dots.* Flash domino card for 3.
▶ Similarly for domino cards from 1 to 4 in random order
▶ Repeat the above sequence using random array cards for 1 to 4.
▶ Repeat the above sequence using domino cards for 1 to 6.
▶ Repeat the above sequence using pairs patterns for 1 to 6.

Purpose, Teaching and Children's Responses

▶ In the case when the card is displayed rather than flashed, and the child is directed not to look at the card, the child might make a part only of the pattern and then look back at the card.

▶ In similar vein to the point made above (see 5.4.1) conceptually based spatio-motor patterns can form a basis for children emerging strategies for adding and subtracting, for example, children will use spatio-motor patterns to keep track of counting in additive or subtractive situations.

5.4.3: Making Auditory Patterns to Match Spatial Patterns

▶ Display domino card for 2. *Clap your hands to show the number of dots on the card.*

▶ Similarly for other domino cards in the range 1 to 6.

▶ *This time I am going to flash the pattern. Look at the pattern and clap your hands to show the number of dots.* Flash domino card for 3.

▶ Similarly for other domino cards in the range 1 to 6.

▶ Repeat the above sequence using random array cards for 1 to 4.

▶ Repeat the above sequence using pairs patterns for 1 to 6.

Purpose, Teaching and Children's Responses

▶ In the case of larger numbers (4, 5, 6) children will attempt to count their auditory pattern. As well, they might alternate rapidly between counting to keep track of the number of claps and visualizing the spatial pattern.

▶ In similar vein to 5.4.1 and 5.4.2 above, conceptually based auditory patterns are also used by children to keep track of counting in additive or subtractive situations.

VOCABULARY

Dots, pattern, clap, flash, array.

MATERIALS

Spatial pattern cards for domino patterns from 1 to 6 (alternatively use dice patterns).
Random array cards for numbers 1 to 4 (several configurations for each number after 1).
Spatial pattern cards for pairs patterns from 1 to 6.

ACKNOWLEDGMENT

These activities were adapted from the work of Robert Wright, and were further developed in the Mathematics Recovery project.

Key Topic 5.5: Finger Patterns

Purpose: To develop initial facility with making finger patterns.

LINKS TO LFIN

Main links: Part C, Finger Patterns.
Other links: FNWS and BNWS, Levels 1–3; SEAL, Stage 1.

TEACHING PROCEDURES

5.5.1: Sequential Patterns for 1 to 5, Fingers Seen

▶ *Watch me use my fingers to make a number.* Raise one finger. *One!* Raise two fingers sequentially. *One, two.* You do that with me. Ready, *One! One, two.* Discuss with the child which hand they are using.
▶ *This time do one, two, three with me.* Ready, *one, two, three.* Similarly with four and five fingers.
▶ *Now use your other hand. Do two with me.* Ready, *one, two! Do three with me.* Ready, *one, two, three!* Similarly with four and five fingers.

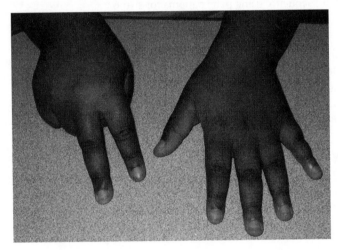

Visible finger patterns

Purpose, Teaching and Children's Responses

▶ There are variations in the extent to which children at this level spontaneously use their fingers in numerical situations. Children who do not seem to use their fingers very much might alternatively focus on counting items available in the classroom environment (for example, blocks on the wall, points on a clock).
▶ Some children at this level need to look at their fingers in order to make finger patterns for the numbers from 1 to 5.
▶ These children might need to count their fingers slowly from one in order to make a pattern corresponding to a given number, or to confirm the numerosity of a finger pattern already made. For these children it can be useful for teachers to model making patterns by sequentially raising fingers.
▶ If the child raises their index finger first, and then their middle finger, they might find it difficult to raise their ring finger.

5.5.2: Sequential Patterns for 1 to 5, Fingers Unseen (Bunny Ears)

▶ *This time don't look at your fingers when you make the number on your fingers. Do three with me. Ready, one, two, three!*
▶ *Now look at your fingers and check to see if you are correct.*
▶ Similarly with two, four and five fingers.
▶ Repeat the previous activity with the child using their other hand. *This time use your other hand to make the patterns and so on.*

Purpose, Teaching and Children's Responses

▶ Children might use their other hand to hold the fingers making the pattern.
▶ In similar vein to points made above (5.5.1) it can be useful for the teacher to model the activity of making finger patterns sequentially without looking at the fingers.

5.5.3: Simultaneous Patterns for 1 to 5, Fingers Seen

▶ *Watch me use my fingers to make a number. This time I am going to raise all of my fingers at once.* Raise two fingers simultaneously. *Two!* Similarly three, four, five. *You do the number I say. Remember, raise all of your fingers at once. Ready, two!* After the child responds, raise two fingers simultaneously. *Two! Compare your finger pattern with mine.*
▶ Similarly three, four and five fingers.
▶ Similarly have children make finger patterns in random order.
▶ Repeat the previous activities with the child using their other hand.

Purpose, Teaching and Children's Responses

▶ Children might use their point finger in their patterns for one and two, and use their other three fingers in their pattern for three.

5.5.4: Simultaneous Patterns for 1 to 5 Fingers Unseen (Bunny Ears)

▶ Place hand on top of head. *Put your hand on your head, like I am doing.* Raise two fingers. *Make a pattern for two like I am doing.*
▶ *This time make a pattern for three.* Similarly for four and five.
▶ Practice making patterns for numbers from one to five in numerical order.
▶ *This time, make the number that I say.* Practice making patterns for numbers from one to five in random order.
▶ Repeat the previous activities with children using their other hand.

Purpose, Teaching and Children's Responses

▶ As indicated in the procedure, initially children can look at the teacher's pattern but cannot see their own pattern (which is on their head). Following this, children are directed to make the pattern without the teacher providing a model.
▶ For numbers after two, children might first raise two fingers simultaneously and then raise other finger(s) sequentially. Similarly, children might first raise three fingers simultaneously to make patterns for four or five.

▶ Related to the previous point, the teacher might ask children to describe how they made the pattern for three, etc.

▶ When several children are doing the activity, children might look at and attempt to copy another child.

5.5.5: Double Patterns for 1 to 5

▶ *Put your hands in front. Make two on your right hand. Make two on your left hand. Count to see how many fingers in all, one, two, three, four! Say after me – two and two make four. This time, look at one hand and then at the other while you say that again, two and two make four.*

▶ Similarly for three and three, four and four, five and five, one and one.

▶ *Now put your hands on your head. Make two and two. Say after me – two and two make four. Now bring your hands down and look at them. Say after me, two and two make four.*

▶ Similarly for three and three, four and four, five and five, one and one.

Purpose, Teaching and Children's Responses

▶ Typically, the first five doubles ($1 + 1, 2 + 2, \ldots 5 + 5$) are among the first additions that the child comes to know. The child's finger patterns constitute enactive instances of these doubles. The term 'enactive' is used here as follows: The child's thinking or knowledge is enactive when it involves action or movement. Thus one could say the child knows '$2 + 2 = 4$' in an enactive sense, in the case where making the finger pattern seems to be an essential part of working out the sum.

▶ Children might make a correct pattern on one hand and an incorrect pattern on the other.

▶ Some children might make the two patterns simultaneously, while others might make one pattern and then make the other.

5.5.6: Using Fingers to Keep Track of Temporal Sequences of Movements

▶ *Watch me as I move my hand.* Move hand in a chopping motion, three times. *One, two, three. Use your fingers to keep track of how many times I move my hand.* Move hand two times. *How many times was that?*

▶ Similarly for one, two, … five, in random order.

▶ Repeat the previous activity with children using their other hand.

Purpose, Teaching and Children's Responses

▶ On these tasks the children's finger patterns are necessarily built sequentially. Of interest is the child's facility to ascribe numerosity to a finger pattern that has been made in this way.

▶ When children build sequential finger patterns in this way (that is, in the context of keeping track of items in temporal sequence, for example, a sequence of movements), some will be able to immediately state the number of the pattern whereas others will need to count their fingers to state the number of the pattern.

▶ The ability to ascribe numerosity to finger patterns which arise sequentially can be an important basis for the activity of keeping track of counts in additive or subtractive situations. For example, in solving a missing addend task such as $8 + x = 11$ (presented using counters), a child might count 'nine, ten, eleven' and raise a finger in coordination with each number word. The child then looks at their three raised fingers and presumably recognizes the finger pattern as symbolizing three. Alternatively, if they are unable to immediately recognize (that is, ascribe a number to) the finger pattern, they might quickly count their fingers from one to three.

5.5.7: Using Fingers to Keep Track of Temporal Sequences of Sounds

▶ *Watch me as I clap.* Make a slow sequence of three claps. *One, two, three. Use your fingers to keep track of how many times I clap.* Make a slow sequence of two claps *How many times was that?*
▶ Similarly for one, two, ... five, in random order.
▶ Repeat the previous activity with children using their other hand.
▶ *Now look away. Use your fingers to keep track of how many times I clap.* Make a slow sequence of two claps.
▶ Similarly for one, two ... five in random order.

Purpose, Teaching and Children's Responses

▶ As in similar vein to the above (5.5.6), some children will be able to state immediately the number of their finger pattern, whereas others will need to count their fingers to do so.
▶ Activities in which children reflect on temporal sequences can form an important basis for keeping track of counting-on or counting-back in additive or subtractive situations.

VOCABULARY

Finger, finger pattern, raise fingers, fingers unseen, at once, hands on head, hands in front, how many, right hand, left hand, keep track, how many, how many altogether.

MATERIALS

None.

ACKNOWLEDGMENT

These activities were adapted from the work of Paul Cobb and colleagues, and were further developed in the Mathematics Recovery project.

Key Topic 5.6: Temporal Patterns and Temporal Sequences
Purpose: To develop facility with copying and counting temporal patterns and temporal sequences

LINKS TO LFIN

Main links: Part C, Temporal Sequences.
Other links: FNWS and BNWS, Levels 1–3; SEAL, Stages 1–2.

TEACHING PROCEDURES

5.6.1: Copying and Counting Temporal Sequences of Movements

▶ *Watch me as I move my hand.* Make six deliberate chopping motions, with a pause before each. *One, two, three, four, five, six. Do that with me. Ready! One, two … six.* Similarly for other numbers in the range 1 to 10.
▶ *This time you count the number of chops I make. Ready!*
▶ *This time, I will say a number and you make that many chops. Ready, four!*
▶ Repeat for numbers in the range 1 to 10.

Purpose, Teaching and Children's Responses

▶ This activity involves coordinating each number word with a simple physical action (chopping). We presume that the child's attention switches back and forth between making the action and saying the corresponding number word. In such situations children are likely to reflect on the number word (or words) previously said. Reflecting on previously said number words in this way can, we believe, strengthen the child's knowledge of the number word sequence and its use in counting.
▶ For larger numbers, children might lose track of counting.
▶ When making the number of chops for a given number (that is, as per the third and fourth point in the teaching procedure), in the case of smaller numbers (for example, 4 or less), children might make a rapid sequence of chops, whereas for larger numbers they will make the sequence more slowly. This relates to children being able to anticipate the sequence of movements prior to commencement in the case of smaller numbers. Thus, in the case of smaller numbers they have an awareness of the complete sequence of movements prior to commencement.
▶ This activity we believe, can lead to the child's development of conceptually based temporal sequences. As before (see Key Topic 5.4), the sequence is said to be conceptually based because the child can enact the sequence (for example, make a rapid sequence of four chops) in the absence of any instance of the sequence which is perceptually available.

5.6.2: Copying and Counting Rhythmic Patterns

▶ *Listen to my pattern and see if you can copy it.* Clap a 2 pattern.
▶ *Now try this one.* Clap a 2–2 pattern.
▶ Similarly with the following patterns: 1–2, 2–1, 1–3, 3–1, 3–3, 2–3, 3–2.
▶ *Try to count how many claps in my pattern.* Clap a 2–2 pattern.
▶ Similarly with the patterns listed above.

Purpose, Teaching and Children's Responses

▶ In the case of longer patterns particularly, children typically will re-present (that is, mentally replay) the pattern, and attempt to count the beats in their represented pattern.
▶ In the case of very short patterns, children will appear to subitize the auditory pattern, that is, they immediately identify the pattern in the sense of correctly ascribing numerosity to it.
▶ As for the previous teaching procedure (5.6.1), these activities, we believe, can lead to the development of conceptually based temporal sequences, which are useful in keeping track of counting in additive or subtractive situations.

5.6.3: Copying and Counting Monotonic Sequences of Sounds

▶ *Try to count how many times I clap.* Make a slow, monotonic sequence of four claps. *How many times?*
▶ Similarly for sequences in the range 1 to 10.
▶ *Now it's your turn. Make four claps.*
▶ Similarly for sequences in the range 1 to 10.

Purpose, Teaching and Children's Responses

▶ Children might attempt to re-present (that is, mentally replay) the sequence after its completion, and to count the re-presented sequence. In the case of longer sequences particularly, children will find this very difficult and subsequently will attempt to count the sequence in real time, that is, as it occurs.
▶ Both kinds of mental activities just described are useful, we believe. Representing the sequence leads to the development of conceptually based temporal sequences (see above). And counting the sequence in real time in similar vein to counting sequences of movements (5.6.1 above), leads to reflection on the previously said number words, and hence a strengthening of the child's knowledge of the number word sequence.

5.6.4: Copying and Counting Arhythmical Sequences

▶ *Try to count how many times I clap.* Make a fast, arhythmical sequence of three claps. *How many times?*
▶ Similarly for sequences in the range 1 to 5.
▶ *Now it's your turn. Make four claps.*
▶ Similarly for sequences in the range 1 to 5.

Purpose, Teaching and Children's Responses

▶ As with the above teaching procedures, these activities have the purpose of making children reflect on temporal sequences and number words sequences.
▶ Children will appear to subitize the shorter sequences (cf. 5.6.2 above).
▶ In the case of longer sequences children might attempt to count the beats of the sequence in real time but are likely to find this difficult. Subsequently children might attempt to re-present the sequence after its completion and count the re-presented sequence. Alternatively children might attempt to immediately ascribe numerosity (that is, to subitize in a sense) or they might attempt to guess the numerosity of the sequence. (cf. 5.6.2 above).

VOCABULARY

Chop, pattern, copy, clap, keep track.

MATERIALS

None.

ACKNOWLEDGMENT

These activities were adapted from the work of Robert Wright, and were further developed in the Mathematics Recovery project.

EXAMPLES OF WHOLE-CLASS LESSONS DESIGNED FOR THE CHILD AT THE EMERGENT STAGE

Lesson 5(a)   Class numeral cards: building the FNWS and identifying numerals to ten and beyond
Lesson 5(b)   Dice patterns: developing knowledge and confidence in the recognition of dice patterns 1 to 6
Lesson 5(c)   Identifying numerals 1 to 10
Lesson 5(d)   Introducing the five frame: linking counting and collections
Lesson 5(e)   Block towers to five: counting and identifying groups of up to five items

LESSON 5(A)

Title:	**Class numeral cards**
Purpose:	To build the forward number word sequence and identify the numerals to ten and beyond ten.
Links to Key Topics:	5.1, 5.2
Materials:	One set of large numeral cards 1–10, child sets of numeral cards 1–10, and playing cards for the Four Kings game.
Introductory Activities:	▶ Saying the number words from one to ten (or beyond).
	▶ Saying the number words from ten to one.
	▶ Saying the number word after. The teacher holds up a large numeral and the class find the number after. If they have a set of numerals they could hold up the appropriate numeral card.
	▶ Repeat the previous task for number before.
Main Focus:	Class numeral track.
	▶ The teacher holds up a stack of numeral cards and tells the class the smallest number is one and the largest is ten.

	▶ The teacher selects a child to choose a card. This child:
	  – shows the class the numeral;
	  – tells the class the number name;
	  – places it on the chalk ledge (or pegs it on a string across the front of the room).

	▶ The teacher asks the class for their opinion of the positioning of the numeral card.
	▶ As each new child is selected to choose a numeral card, they need to place it in relation to those numerals already displayed. If children have difficulty reading or placing a card, they can choose another child to assist.

▶ When all of the cards are placed the teacher:
  − points to various cards and asks the class to state the number name;
  − asks the class to say sequences of numbers both forward and backward with the teacher pointing to each numeral card;
  − turns over individual cards and asks the class to tell the number that is hidden;
  − turns over several cards and then points to these one at a time asking the children to state the number;
  − turns over all but one of the cards and repeats the previous task.

Group/Individual Tasks: The Four Kings.

The purpose of this activity is to assist children in identifying numerals and with forward and backward number sequences.

▶ Participants: Individual or pairs.

▶ Equipment: Pack of cards (take out the queens and jacks). This will leave 4 suits including the kings (a total of 44 cards).

▶ Explanation.
  − Cards are laid out face down in rows of ten.
  − The four left over are put to one side.
  − The player begins by turning over one of the four extra cards and placing it in the correct position after taking the card that is already in that position. This card is then put into its correct position.
  − When a king is turned up another card is selected from those at the side.
  − The game proceeds until the four kings are turned up.
  − The object of the game is to turn over as many cards as possible before the four kings are turned over.

Note: It is useful at the start to mark each row with the appropriate symbol (hearts, diamonds, clubs, spades).

▶ Modification
  (1) Instead of using traditional playing cards, make up your own cards from four different colored cardboards.
  (2) Have cards with different sets of numerals (for example, from 10 to 19 or 45 to 57, and so on).
  (3) A simpler game involves only two rows of cards and two kings (even two rows with blue and pink cards 1 to 5) would be useful for some children.

Conclusion/Summary: The class is assembled together in a group.

▶ The teacher gives tasks such as:
  − start counting from one;
  − start counting from three;
  − count backwards from seven.

▶ The teacher notes those children who are confident and those who need further assistance with:
  − identifying the numerals and saying the number word sequences forward and backward from different starting points.

LESSON 5(B)

Title:	**Dice patterns**
Purpose:	Children develop knowledge and confidence in recognizing the dice patterns 1 to 6.
Links to Key Topics:	5.4, 5.2
Materials:	Dice pattern transparencies or large cards for class display, dominoes, child dice pattern cards or dice.

Introductory Activities:

▶ Flash dice patterns (1–6). Have children tell how many dots are on the card and hold up the appropriate number of fingers.

▶ Teacher holds up a numeral and children respond by holding up the appropriate dot card or show the correct face on a dice. (If cards are used each child or pair has a set of dice pattern cards.)

Main Focus: Dice pattern Bingo.

▶ Each child or pair of children is given a Bingo board (see diagram) containing the numerals 1 to 6.

6	1	4	2	3
4	6	5	5	1
3	2	3	4	6

▶ The teacher or leader rolls a die and shows the class the pattern that was face-up.

▶ Alternatively, a well-shuffled set of large dice pattern cards can be used, with the teacher displaying the top card to the class.

▶ If children have this numeral on their board, they place a counter on this square.

▶ If there is more than one square with this numeral they choose on which square they will place their counter.

▶ When a child has three counters in a row (or a larger number of counters if appropriate boards have been made) this child calls out 'Bingo!'

▶ As a variation, the board can have the patterns and the teacher calls out numbers, which are generated by throwing a die.

Group/IndividualTasks: Dice and numeral dominoes.

Pairs of children play dominoes with conventional pieces or with specially constructed domino and numeral cards as shown in the diagram.

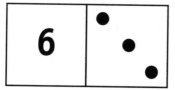 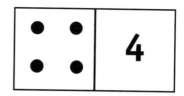

Conclusion/Summary:	With the class (or group) assembled:

- the teacher asks children to describe the dice patterns;
- some more dice patterns are flashed and children asked to tell the number of dots.

LESSON 5(C)

Title:	**Identifying numerals**
Purpose:	To assist children to learn the numerals 1 to 10.
Links to Key Topics:	5.2
Materials:	Playing cards or numeral cards constructed by teacher.
Introductory Activities:	▶ Number words to twelve.
	▶ Number words in conjunction with the numerals on the numeral track. Teacher drags the numeral roll from under a screen revealing the next numeral. Sometimes displaying the numeral before the children say the number word and sometimes after.
Main Focus:	Bingo using numerals from 1 to 10.
	– The traditional game of Bingo is played with the whole class or group (that is, those that need experience with the numerals 1 to 10).
	– The game can also be played with pairs (one child who knows the numerals teamed with a less confident child).
Group/Individual Tasks:	Fish.

The game of Fish will complement Bingo and help children become more confident with the numerals to 10.

- ▶ Pairs or groups of three can be arranged with players at similar levels or a more advanced child could be paired with a child needing more experience with numerals to 10.
- ▶ Instructions:
- ▶ Five cards are dealt to each player.
- ▶ These are put down in each player's hidden space (each player has a hidden screen because young children have difficulty holding cards in their hand).
- ▶ The object is to make pairs of numbers (the kings, queens and jacks can be taken out if the teacher so desires) and the stack of cards is placed face down.
- ▶ The players have a turn to ask another child if they have a certain card (which would give them a pair).

▶ If the child to whom the request has been made has the card, then this is given to the player who places the pair face-up on the table.

▶ If the child to whom the request has been made does not have the required card, the player is told to FISH, which means the top card from the stack is selected.

▶ The game proceeds until one player runs out of cards.

▶ Players can count up their pairs.

Conclusion/Summary: The class is assembled and the teacher chooses children in turn to:

– select a numeral card from a randomly arranged stack and state the number on it;

– find the correct numeral card after a number has been stated.

LESSON 5(D)

Title: **Linking counting and collections**

Purpose: To introduce children to the five frame and use it to build combinations to five.

Links to Key Topics: 5.1, 5.2, 5.3, 5.4

Materials: Transparent counters, five frames for the overhead projector, child five frames (5 for each child or pair of children); five red and five blue counters for each child or pair of children.

Introductory Activities: ▶ *Show me two fingers. Show me five fingers. Show me three fingers.*
– Extend the children to just beyond their limit or comfort zone.
– Those children who are not as advanced will be assisted by having the models of more advanced children. They could also be seated close to more advanced children.

▶ On the overhead projector display a group of three red counters and a group of two blue counters. *How many red counters are there? How many blue counters ore there? How many counters are there altogether?*

 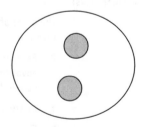

Main Focus: Packing chocolates.

– In this task the five frame is placed on an overhead projector.

– Display a five space chocolate box. *How many spaces are there? Let's count all of the spaces.*

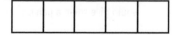

▷ Place three red transparent counters in the frame. *How many red counters are there? How many empty spaces are there?*

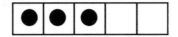

▷ Now place two blue counters in the frame. *How many red counters are there? How many blue counters are there? How many counters are there altogether?*

▷ Give other similar tasks using red and blue counters in the five frame.

Group/Individual Tasks: ▷ Select children to come out to the overhead projector and place given numbers of counters in the five frame. *Who could come out and put four red counters in the box? Can anyone tell me how many empty spaces there will be?*

▷ Finding different combinations of red and blue counters. Each child or pair of children is given five chocolate boxes (five frames), five red counters and five blue counters. *Use your chocolate boxes to make as many different red and blue chocolates. The rule is that the reds must be next to each other and the blues must be next to each other You can't have this mixture.* Display this on the overhead projector.

Conclusion/Summary: – Discuss the results of the red and blue combinations.

  – Have children make different combinations on the overhead projector.

LESSON 5(E)

Title:	**Block towers to five**
Purpose:	To help children count groups of up to five items and to identify groups of up to five items.
Links to Key Topics:	5.3, 5.4
Materials:	Blocks (for example, Unifix – blocks that attach).
Introductory Activities:	▷ Saying the number words from one to ten (or beyond).

▷ Saying the number words from ten to one.

▷ Screen a numeral track (1 to 10) and have the class say the number word as each numeral is revealed. It may be more appropriate to use the numerals 1 to 5. This needs to be decided by the teacher.

Main Focus:

▶ Each child or pair of children has a set of 15 blocks (preferably of the one colour).

▶ Teacher says:
  – *Make a tower 1 block high.*
  – *Make a tower 2 blocks high.*
  – *Make a tower 3 blocks high.*
  – *Make a tower 4 blocks high.*
  – *Make a tower 5 blocks high.*

▶ Teacher says:
  – *Hold up the 2 tower.*
  – *Hold up the 5 tower.* And so on.

▶ Teacher briefly displays a tower and the children hold up the corresponding number of fingers. This is repeated for all the towers to 5.

▶ Teacher gives tasks like:
  – *Make a 2 tower. How many more blocks do I need to make a 5 tower?*
  – *Make a 4 tower. How many blocks do I need to take off to have a 3 tower?*

Group/Individual Tasks: Tower and Numeral Snap.

Pairs of children play the game of Snap using tower and numeral cards.

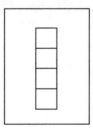 

Conclusion/Summary: With the whole class the teacher assesses:

– knowledge of identifying numerals to 5;

– identifying towers to 5;

– making towers to 5.

6
Teaching the Perceptual Child

This chapter focusses on teaching children at the Perceptual Stage on the SEAL (Stages of Early Arithmetical Learning – see Chapter 1). First, we provide a detailed summary of the knowledge and strategies typical of this stage. This is followed by descriptions of six key topics which can provide a basis for teaching the child at the Perceptual Stage. The six key topics consist of a total of 32 teaching procedures.

THE TYPICAL PERCEPTUAL CHILD

This section provides an overview of a typical child at the Perceptual Stage, that is, at Stage 1 of the Stages of Early Arithmetical Learning. Table 6.1 sets out the stage and indicative levels for the perceptual child on the models pertaining to FNWSs, BNWSs, Numeral Identification and Tens and Ones. The overview discusses the first four aspects of early number knowledge listed in Table 6.1, and also discusses knowledge of spatial patterns and facility with finger patterns.

Table 6.1 Stage and Levels of a Typical Perceptual Child

Model	Stage/level
Stage of Early Arithmetical Learning (SEAL)	1
Level of Forward Number Word Sequences (FNWSs)	3
Level of Backward Number Word Sequences (BNWSs)	1
Level of Numeral Identification	1
Level of Tens and Ones Knowledge	0

Stage of Early Arithmetical Learning

The child at the perceptual stage is able to count a collection of counters, for example 13 or 18 counters, but is not able to solve additive tasks involving two collections in cases where one or both of the collections are screened (the child is told how many counters are in the screened collection(s)). Some children at this stage seem to have difficulty establishing the numerosity of two unscreened collections taken together. In the case of a task such as establishing the numerosity of a single collection, for example a collection of 18 counters, the child can correctly count the counters by ones from one to 18. In the case of two separate collections, for example a collection of 9 red counters and a collection of 6 blue counters, some perceptual children do not seem to alternatively regard the two collections as one for the purposes of establishing the overall numerosity. This difficulty has been observed to arise even in cases involving very small numbers of counters (for example, five and two). Thus when asked how many counters in all, the child answers 'five' or 'two', but does not seem to realize that they can make one count from one, to count the counters in both collections. Also, in the situation just described (that is, a collection of 5 counters and a collection of 2 counters – labeled verbally with the number words 'five' and 'two' respectively), some children seem to regard the numbers 5 and 2, as constituting a 2-digit number, that is, '52'. The latter behavior can result from premature teaching of place value ideas.

FNWSs and BNWSs

The child at the Perceptual Stage typically has good facility with FNWSs in the range 1 to 10. In addition, the child might be able to say the FNWSs into the twenties, sometimes stopping at 29, or might say the number word sequence beyond 30. Nevertheless the child might well have difficulty with tasks involving saying immediately the number word after a number beyond 10 (for example, say the number word after 13). Thus the child might use a dropping back strategy or not be able to solve the task at all. In the case of BNWSs the child typically will be able to say the number words from 10 to 1, but might have difficulty with tasks involving saying immediately the number word before a number in the range one to ten. Again, the child might not be able to solve the task at all, or might use a dropping back strategy.

Numeral Identification

Children at this stage typically can recognize and identify numerals in the range 1 to 10, but might not be able to do so with numerals in the teens. They might incorrectly identify '12' as 'twenty' or 'twenty-one', or a numeral in the range '13' to '19' as the corresponding decade number (for example, '17' is named as 'seventy'). If asked to write a teen numeral, for example '16', the child might first write the digit '6', and then write the digit '1' to the left of the '6'. The likely explanation for this is that, rather than form a mental image of the numeral prior to writing, the child determines each of the digits '6' and '1' from the sound image of the number word (that is, 'six – teen'), first writes the digit '6', and only then realizes that the digit '1' should be placed to the left of the digit '6'.

Spatial Patterns and Finger Patterns

Children at this stage typically can subitize in the range one to four, that is, they can correctly ascribe number to spatial configurations of dots, particularly when the configuration forms a pattern (such as the patterns on dice) in the range one to four, and typically do not count from one when doing so. Some children at this stage have facile finger patterns for numbers in the range 1 to 5, and might use their finger patterns to solve additive tasks in cases where both numbers are in the range 1 to 5. This frequently occurring strategy involves making a finger pattern on one hand to correspond with the first addend and, similarly, making a finger pattern on the other hand for the second addend. The child then counts their raised fingers from one to obtain the answer. Establishing the finger patterns might involve sequentially raising fingers in coordination with counting to the number of the corresponding addend. Children with more facile finger patterns will raise fingers simultaneously to correspond to an addend, and do not count from one when doing so. At the same time these children typically will not have facile finger patterns for the numbers in the range 6 to 10. Thus if asked to show 7 on their fingers they will raise seven fingers sequentially while counting from one to seven.

The Way Forward

The perceptual child has a significant body of early number knowledge which can be strengthened and extended with appropriate instruction. Facility with FNWSs can be consolidated in the range 1 to 10 and extended in the range 1 to 30 and beyond, and facility with BNWSs can also be consolidated and extended. Children's facility with counting to solve additive tasks can be extended using settings involving screened collections and settings involving rows of counters some of which are screened. Knowledge of numerals and numeral sequences can be extended into the teens, that is, children should learn to name the teen numbers and become familiar with the numeral sequence to 20. Children can become more facile with naming spatial patterns based on grids with two rows, for numbers up to 10, and can begin to think conceptually about spatial patterns (for example, a 6 pattern can alternatively be regarded as a 4 pattern and a 2 pattern). Finger patterns for numbers up to 5 can be consolidated and children can learn to use a finger pattern for five to build finger patterns for numbers in the range 6 to 10. As well, children can learn to build finger patterns for the even numbers (2, 4, ... 10) consisting of the same number of fingers on each hand (for example, 3 and 3). Finally, children should develop initial knowledge of equal groups and sharing (that is, as a basis for multiplication and division).

The next section contains six key topics and 32 teaching procedures which form the basis of an appropriate instructional programme for the perceptual child.

Key Topic 6.1: Number Word Sequences from 1 to 30

Purpose: To develop knowledge of number word sequences in the range 1 to 30.

LINKS TO LFIN

Main links: FNWS and BNWS, Levels 1–4.
Other links: SEAL, Stage 1.

TEACHING PROCEDURES

6.1.1: Saying Short FNWSs

▶ *Start from six and count up to 10. Now start from 8 and count up to 15.*
▶ *Start from 20 and count up to 25.*
▶ *Similarly, 18 to 24, 22 to 30, and so on.*

Purpose, Teaching and Children's Responses

▶ The activities in this key topic focus on number word sequences alone, that is, not in conjunction with numerals, or collections or rows of items. For children at this level we believe it is important to include these activities as well as the activities in Key Topic 6.2, which focus on numerals in conjunction with number words, and those in Key Topic 6.3, which focus on using number words in the context of counting the items in collections or rows. Thus this variety of activities contributes to the overall development of the child's number knowledge, and each particular kind of activity is important for this development.

▶ Give children sufficient time to think hard and to reflect carefully on their responses.

▶ Children might have difficulty saying FNWSs that do not start from one.

▶ Children might have difficulty in pronouncing 'thirteen', 'fourteen', … 'nineteen' and might confuse these with decade names ('thirty' and so on).

▶ Children might not know the number word after 29 or might say 'twenty-ten'.

▶ Children might omit particular number words (for example, 'thirteen', 'twenty').

6.1.2: Saying Short BNWSs

▶ *Start from 10 and count back to 7. Now start from 12 and count back to 8.*

▶ *Start from 20 and count back to 16.*

▶ *Similarly, 24 back to 20, 30 back to 26, and so on.*

Purpose, Teaching and Children's Responses

▶ BNWSs in the teens are usually much more difficult for children than FNWSs in the teens.

▶ Children quite frequently use a dropping back strategy (subvocally or aloud) to determine a particular number word, for example, the child says the FNWS from ten to figure out what comes before 'thirteen'. This should occur less frequently as the child's knowledge of number word sequences strengthens.

▶ Some children might consistently omit a particular number word, for example 'thirteen' or 'twenty'.

6.1.3: Saying Alternate Number Words Forwards and Backwards

▶ *Let's take turns to say the numbers. I will say one, then you say two, then I will say three, then you say four, and we will keep going like that. Ready, one, (two), three, (four), …* Continue to about twenty.

▶ *This time you start with one. Ready, (one), two, (three), four, …* to about twenty.

▶ *This time you start from ten. Ready, (ten), eleven, (twelve), thirteen, …* Continue into thirties.

▶ *Let's try that going backwards. Let's go backwards from twelve. I'll start off. Ready, twelve, (eleven), ten, …*

▶ *This time we will go backwards and you can start from thirty. Ready, (thirty), twenty-nine, (twenty-eight), …*

▶ Similarly with other FNWSs and BNWSs into the thirties.

Purpose, Teaching and Children's Responses

▶ As indicated in Chapter 5 (see notes for Key Topic 5.3), this kind of activity requires a distinctive process of reflection by the child.

▶ Reflecting on number words spoken by the teacher and the child can serve to strengthen knowledge of number word sequences.

6.1.4: Saying the Next One, Two or Three Number Words Forwards

▶ *I'm going to count from a number, and I want you to say the next number after I stop. Ready, ten, eleven, twelve —. Ready, eighteen, nineteen, twenty, twenty-one, —.*

▶ *I'm going to count from a number, and I want you to say the next two numbers after I stop. Ready, ten, eleven, twelve —. Ready, eighteen, nineteen, twenty, twenty-one, —.*

▶ *I'm going to count from a number, and I want you to say the next three numbers after I stop. Ready, one, two, three, four —. Ready, nine, ten, eleven, twelve, —. Ready, twenty-one, twenty-two, twenty-three, —.*

Purpose, Teaching and Children's Responses

▶ Facility with saying one, two or three number words forwards, forms an important basis for counting-on strategies (counting-up-from and counting- up-to).

▶ Children quite frequently use a dropping back strategy (subvocally or aloud) to determine a particular number word, for example, the child says the FNWS from ten to figure out what comes after 'fourteen'. This should occur less frequently as the child's knowledge of number word sequences strengthens.

6.1.5: Saying the Next One, Two or Three Number Words Backwards

▶ *I'm going to count backwards, and I want you to say the next number backwards, after I stop. Ready, twelve, eleven, ten, —. Ready, eighteen, seventeen, sixteen, —. Ready, thirty, twenty-nine, twenty-eight, —.*

▶ *I'm going to count backwards, and I want you to say the next two numbers backwards, after I stop. Ready, eleven, ten, nine, —. Ready, eighteen, seventeen, sixteen, —. Ready, twenty-six, twenty-five, twenty-four, —.*

▶ *I'm going to count backwards, and I want you to say the next three numbers backwards, after I stop. Ready, ten, nine, eight, seven, —. Ready, twenty, nineteen, eighteen, —. Ready, twenty-four, twenty-three, twenty-two, twenty-one, —.*

Purpose, Teaching and Children's Responses

▶ Facility with saying one, two or three number words backwards, forms an important basis for counting down strategies (counting-down-from and counting-down-to).

▶ Children frequently use a dropping back strategy (similar to 6.1.2 above).

▶ Children are likely to encounter more difficulty when fewer number words are said by the teacher. Thus 'seventeen, sixteen, —,' is likely to be more difficult than 'nineteen, eighteen, seventeen, sixteen, —'.

6.1.6: Saying the Number Word After

▶ *I'm going to say a number and I want you to say the number just after the one I say. Ready, seven! Four! Eight!*

▶ Use number words in the range 1 to 30.

Purpose, Teaching and Children's Responses

▶ As for Key Topic 5.1.

6.1.7: Saying the Number Word Before

▶ *I'm going to say a number and I want you to say the number before the one I say. Ready, four! Ten! Three!*

▶ Use number words in the range 1 to 30.

Purpose, Teaching and Children's Responses

▶   As for Key Topic 5.1.

VOCABULARY

One, two, three … thirty, count from, count-up-to, count-back-to, next number, next two numbers, next three numbers, backwards two/three numbers, number of jumps. See also Key Topic 5.1.

MATERIALS

None.

ACKNOWLEDGMENT

These activities were adapted from the work of Robert Wright, and were further developed in the Mathematics Recovery project.

Key Topic 6.2: Numerals from 1 to 20

Purpose: To develop knowledge of numerals and numeral sequences in the range 1 to 20.

LINKS TO LFIN

Main links: Numeral Identification, Levels 1–2.
Other links: FNWS and BNWS, Levels 1–4; SEAL, Stages 1–2.

TEACHING PROCEDURES

6.2.1: Numeral Sequences, Forwards and Backwards

▶   Place out the numeral sequence from 11 to 15 (use one strip of numerals or 5 cards). *Watch me and listen as I count forwards and backwards.* Point to each numeral in turn, while counting from 11 to 15, and then from 15 to 11. *Eleven, twelve, … fifteen. Fifteen, fourteen, … eleven.*
▶   Similarly with the numeral sequence from 16 to 20, and 11 to 20.

Purpose, Teaching and Children's Responses

▶   Children's knowledge of the corresponding FNWS facilitates the activity of reading a numeral sequence forwards. Thus a child might appear to be reading the numerals from 11 to 15 in turn when their response is based mainly on their knowledge of the sequence of number words from eleven to fifteen.

Numeral sequences, forwards and backwards

▶ Notwithstanding the point in the previous sentence, these activities serve to strengthen children's knowledge of numerals, numeral sequences and number word sequences.

▶ In the case of reading a numeral sequence backwards, knowledge of the corresponding BNWS might facilitate reading of the sequence backwards, but this is likely to happen to a much lesser extent (compared with reading a sequence forwards) because children's knowledge of the corresponding BNWS is likely to be less facile than their knowledge of the FNWS.

▶ As before, these activities serve to strengthen knowledge of numerals, numeral sequences, number words and, in particular, numeral sequences backwards and BNWSs.

6.2.2: Sequencing Numerals

▶ Place out the cards from 11 to 15, randomly arranged. *Put these cards in order*. Direct the child to arrange cards left to right in increasing order. *Now say the numbers as you point to them*.

▶ Similarly order the cards from 16 to 20 and 11 to 20.

Purpose, Teaching and Children's Responses

▶ Some children might be able to arrange a set of numerals in the correct sequence (for example, the numerals from 11 to 15) without necessarily being able to name all of the numerals. This might occur, for example, if the child has had relatively extensive experience with numeral sequences without necessarily having experience with the corresponding number word sequences, for example, the numeral sequence from 1 to 10 or 1 to 20 might be fixed on the child's desk or on a wall in their classroom.

6.2.3: Numeral Recognition

▶ Place out the cards from 11 to 15, randomly arranged. *Point to the number 12. Point to the number 13. Point to the number 11*.

▶ Similarly using the cards from 16 to 20, 11 to 20 and 1 to 20.

Purpose, Teaching and Children's Responses

▶ Some children find recognition of particular numerals easier than identification of those numerals. For example, some children might recognize '12' among the numerals from 11 to 20 (randomly arranged) but have difficulty in identifying '12'. Thus some children have particular difficulty in generating the sound image 'twelve' upon seeing the numeral '12'. (See Chapter 1, LFIN Part B.)

6.2.4: Numeral Identification

▶ Place out the cards from 11 to 15, randomly arranged. Point to the card for 12. *What number is this?* Point to the card for 16. *What number is this?*
▶ Similarly using the cards from 16 to 20, 11 to 20 and 1 to 20.

Purpose, Teaching and Children's Responses

▶ For the numerals from 13 to 19, some children seem to read each of the digits in turn, commencing with the right hand digit. Thus 19, is named for example, by looking first at the '9' and saying 'nine', and then looking at the '1' and saying 'teen'.
▶ Many children have particular difficulty with identifying '12'. For many children for example, identifying '11' seems much easier than identifying '12'. '12' is often named as 'twenty' or 'twenty-one'. This might occur frequently with children who read the numerals from 13 to 19 by first reading the right-hand digit.
▶ It is not unusual for children in their regular classroom situations to use available numeral sequences to identify numerals. Consider a scenario where the numeral sequence from 1 to 20 is fixed onto a child's desk or appears on the classroom wall (cf. 5.2.1 above). When confronted with the task of identifying the numeral '12' (that is, generating the number word 'twelve' upon seeing '12') a child might count subvocally (and typically quite quickly) along the (visibly available) numeral sequence from one.
▶ It is important we believe, that teaching should directly counter the strategy described in the previous point. This can be done by making the necessary changes to the instructional environment, and using instructional strategies that promote generating the names of numerals immediately rather than by saying a number word sequence forwards.
▶ Thus, although for most children, activities with numeral sequences (see 5.2.1 and 5.2.2 above) constitute one important path for learning about numerals, for some children, it might be necessary to cease instruction involving numeral sequences in order for the child to develop the strategy of immediately generating names of numerals.
▶ Children can and should learn the names of 2-digit numerals at a time when they might have little understanding of the place value associated with the numerals (that is, understanding that the left-hand digit stands for tens).
▶ As in Key Topic 5.2, some children might frequently confuse particular numerals. In such cases, visual discrimination tasks can be useful. For example, use a pile of cards where either '12' or '20' appears on each card (or '12' and '21' as necessary). The child's task is to sort the cards into two piles (12s and 20s).

6.2.5: Numeral Tracks – 11 to 15, 16 to 20 and 11 to 20

▶ Place out the numeral track from 11 to 15, with numerals uncovered. *Watch me as I count forwards and backwards.* Point to each numeral in turn, while counting from eleven to fifteen, and then fifteen to eleven. *Now you count forwards and backwards and point to each number in turn.*

▶ Close lids on the numeral track and repeat previous activity, uncovering the lids after saying each number.

▶ Place out the numeral track from 11 to 15, with all numerals covered. Uncover the numeral 13. *What number is that?* Leave the numeral 13 uncovered, and point to the lid covering the numeral 14. *What number is that?* Direct the child to uncover the numeral 14. *Were you correct?*

▶ Similarly for other numerals from 11 to 15.

▶ Repeat the previous activities using the numeral tracks from 16 to 20, 11 to 20 and 1 to 20.

Before and after

Purpose, Teaching and Children's Responses

▶ With frequent use of the numeral track children will learn to associate the name of the numeral under a given lid, with the position of the lid, for example, the child will know that the second last lid for the numeral track from 11 to 20, covers the numeral 19.

▶ Activities with numeral tracks, we believe, facilitate the development of visualized images of numerals. Typically these images are developed in conjunction with the corresponding number words, that is, the verbally based knowledge involving the names of the numerals.

▶ Extensive work with numeral tracks and other settings involving numeral sequences can, we believe, facilitate the development of knowledge of numerals which is not necessarily subsequent or secondary to the development of knowledge of number words. We have observed, for example, children who, when attempting to solve a problem, appear to reason with visualized images of numerals rather than with number words – the child answers by saying 'it's got a six and an eight' rather than by saying 'sixty-eight'.

6.2.6: Numeral Rolls 1 to 20, 1 to 50

▶ Display the numeral roll from 1 to 20. *This is called a numeral roll. It shows the numbers from 1 to 20 in order. You say the numbers with me as I unroll it. Ready, one, two, three, — twenty. Now let's say the numbers backwards. Ready, twenty, nineteen, — one.*

▶ Display the numeral roll from 1 to 50. *This numeral roll shows the numbers from 1 to 50 in order. You say the numbers with me as I unroll it. Ready, one, two, three, — twenty. Let's try to go further than twenty. Twenty-one, twenty-two, and so on. Now let's say the numbers backwards from twenty-five, and so on.*

▶ Using the numeral roll from 1 to 50 point to each set of 10 numerals in turn. *These numbers are called the ones. These numbers are called the teens. These numbers are called the twenties, and so on. Point to the twenties. Point to the forties, and so on.*

Purpose, Teaching and Children's Responses

▶ A slotted card can be used in conjunction with the numeral roll to display one numeral at a time (make two vertical slots in the card and thread the roll through both slots so that only one numeral is displayed).

▶ We believe activities with numeral rolls can help to develop children's knowledge in distinctive ways. For example, these activities can lead to the development of a broad sense of order (for example, the eighties are after the twenties) and sequence of number words and numerals. Thus teaching procedures based on the numeral roll are used before teaching procedures based on the hundred square (Key Topic 7.2). In the case of the hundred square, the sequence of decades is less apparent than in the case of the numeral roll.

▶ Children will use their knowledge of number word sequences to read the numerals on the numeral roll.

▶ After the child is familiar with the numeral roll, numeral rolls with color coding by 10s (or 5s) can be used. Thus, in the range 1 to 10, the numeral roll is green, red in the range 11 to 20, green in the range 21 to 30, and so on.

VOCABULARY

Numeral roll, teens, twenties, thirties, and so on. See also Key Topic 5.2.

MATERIALS

Numeral sequences from 11 to 15, 16 to 20, 11 to 20, 1 to 20.
Multiple copies of numeral cards (or plastic numerals) for each number in the range 1 to 20.
Numeral track from 11 to 15, 16 to 20, 11 to 20, 1 to 20.
Numeral rolls from 1 to 20 and 1 to 50.

ACKNOWLEDGMENT

These activities were adapted from the work of Robert Wright, and were further developed in the Mathematics Recovery project. The numeral track was developed by Garry Bell (Southern Cross University).

Key Topic 6.3 Figurative Counting

Purpose: To develop figurative counting strategies.

LINKS TO LFIN

Main links: SEAL, Stage 2.
Other links: FNWS and BNWS, Levels 1–4.

TEACHING PROCEDURES

6.3.1: Counting Items in Two Collections, with First Collection Screened

▶   Place out 5 red counters. *Here are five red counters. I'm going to cover those five counters.* Place a screen over the five red counters. *Here are two green counters. Five and two, how many counters are there in all?*

▶   Similarly, with 6 red and 3 green, 10 red and 2 green, and so on.

Purpose, Teaching and Children's Responses

▶   The sizes of the numbers in the first and second collections should be chosen carefully. Initially the first number is in the range 4 to 12. This can be extended beyond 12 virtually as soon as the child has a reasonable knowledge of the corresponding FNWSs. Thus the range of size of the first number may be extended to the 20s and beyond, but continue to use numbers in the ones and teens as well as higher numbers. The second number is typically 2, 3 or 4, or possibly 5 or 6. The second number is limited to this narrow range because that is the range in which it is considered useful for children to become facile at keeping track of counting. Generally, the higher the second number, the more difficult the task.

▶   Children will count from one to count the first screened collection, rather than count-on from the number in the first collection.

▶   When counting from one to count the counters in the first collection, children might make a sequence of pointing actions over the screen.

▶   If the child does not seem able to solve the task, the teacher can model all or part of the solution.

▶   When the child has solved the task, remove the screen and direct the child to check.

6.3.2: Counting Items in Two Collections, with Second Collection Screened

▶   Place out 6 red counters. *Here are six red counters.* Place out 3 green counters. *Here are three green counters. I'm going to cover the three green counters.* Place a screen over the 3 green counters. *How many counters are there in all?*

▶   Similarly, with 8 red and 3 green, 10 red and 4 green, and so on.

Purpose, Teaching and Children's Responses

▶ The general points above (6.3.1) apply here as well (concerning number size, checking, and so on).
▶ Children will count the counters in the visible collection from one, and then continue to count the screened collection.
▶ There is a range of strategies that children use to keep track of the counters in the screened collection. For example: children might apparently visualize the items in the screened collection; children might make a sequence of movements corresponding in number to the number of counters in the second collection; and children might keep track of the number of times they count – thus they might keep track of three counts when saying 'seven, eight, nine'.
▶ The last mentioned strategy might involve explicitly double counting ('seven is one, eight is two, nine is three').
▶ Some children find it difficult to conceive of two collections alternatively as one collection for the purpose of counting how many counters in all. This might involve answering 'six, three', when asked how many counters in all. In this case the teacher should directly teach the idea of two collections being combined for the purposes of counting the items in both collections.

6.3.3: Counting Items in Two Screened Collections

▶ Place out 8 red counters. *Here are eight red counters. I'm going to cover those eight counters.* Place out 2 green counters. *Here are two green counters. I'm going to cover the two green counters.* Place a screen over the 2 green counters. *How many counters are there in all?*
▶ Similarly, with 5 red and 4 green, 9 red and 3 green, and so on.

Purpose, Teaching and Children's Responses

▶ The notes for 6.3.1 and 6.3.2 apply similarly to 6.3.3.

6.3.4: Counting Items in a Row with Some Items Screened

▶ Place out a row of 20 dots. *Count the dots from one, forwards and backwards.*
▶ Place a marker on or adjacent to the sixth dot. *This is number six.* Place a small screen over the seventh and eighth dots. *There are two under here.* Point to the screen. Place a different colored marker on the ninth dot. *What number is this one?*
▶ Similarly, with seventh and eleventh dots, twelfth and fifteenth dots, and so on.

Purpose, Teaching and Children's Responses

▶ The note for 6.3.1 about sizes of the numbers (see above) applies here as well.
▶ These kinds of tasks can be particularly useful for engendering counting-on and counting-back because of the good likelihood that the child becomes aware of, for example, associating the number word 'six' with the sixth counter, without having to first count from one to six. Thus the child might solve these tasks without counting from one.
▶ Children might incorrectly state the number of the last screened dot, as the number of the next, that is, unscreened dot.

VOCABULARY

Counters, group, collection, how many, altogether, cover, number, count, row, dots, forwards, backwards, red counters, green counters.

MATERIALS

Counters of two colors.
Row of 20 dots.
Small cubes of two colors to use as markers.
Small screens to cover 1–4 dots in a row.
Two screens each consisting of a rectangular sheet of cardboard.

ACKNOWLEDGMENT

These activities were adapted from the work of Leslie Steffe and the work of Robert Wright.

Key Topic 6.4: Spatial Patterns

Purpose: To further develop facility to ascribe number to regular spatial patterns.

LINKS TO LFIN

Main links: Part C, Spatial Patterns.
Other links: Part C, Combining and Partitioning.
FNWS and BNWS, Levels 1–3; SEAL, Stage 1–2.

TEACHING PROCEDURES

6.4.1: Partitioning Visible Patterns to 6

▶ Display a Domino-4 pattern. *How many dots do you see? Can you point to 2 dots in the pattern? Can you point to another two dots? How many lots of two can you see?*
▶ Display a Domino-5 pattern. *How many dots do you see? Can you point to 2 dots in the pattern? Can you point to another 2 dots? How many lots of 2 can you see? What else can you see?*
▶ Similarly using domino-6, domino-3 and pairs patterns for 3 to 6.

Purpose, Teaching and Children's Responses

▶ The activities in this key topic have the purpose of making children think about and reflect on small numbers from the perspective of a group rather than a composite. Thus, for example, two is regarded as a group or unit of two rather than a composite consisting of two ones.

▶ In similar vein, children might come to regard a group of six as consisting of three groups of two, and so on.

▶ Children might have a preference for counting dots from one, for example, they might say 'one, two', to refer to a group of two dots, rather than saying 'two'. This is likely to work against their regard for numbers as groups rather than composites. In this case the teacher can model how the spoken number word 'two' for example, can be used to label a group of two dots.

6.4.2: Partitioning Flashed Patterns to 6

▶ Flash a domino-4 pattern. *How many dots did you see? Can you describe it another way?*
▶ Similarly using domino and pairs patterns for 3 to 6.

Purpose, Teaching and Children's Responses

▶ The procedure of flashing the patterns has the purpose of developing children's conceptually based visualized images for the numbers from one to ten, and developing children's ability to use these images to combine and partition numbers in the range 1 to 10.

▶ Children might attempt to count the dots while the pattern is being displayed. Alternatively, children might immediately generate a number word (that is, corresponding to the number of dots) or might attempt to visualize the pattern, and count the dots corresponding to their visualized pattern.

▶ As for 6.4.1 above, children should be encouraged to say the corresponding number word immediately, rather than attempt to count from one.

6.4.3: Partitioning Visible Patterns to 10

▶ Display a pairs-8 pattern. *How many dots do you see? Point to 4 dots in the pattern? Point to another 4 dots. How many lots of 4 can you see? Make a pattern in the air to match the 4 dots.*
▶ *Point to 2 dots in the pattern. Point to another 2 dots. How many 2s can you see? Try to find other numbers in the pattern. Make a pattern in the air to match the 4 dots.*
▶ Similarly using pairs patterns for 7, 9 and 10.

Purpose, Teaching and Children's Responses

▶ As for 6.4.1 above.
▶ An important goal is for children to be able to recognize the pairs patterns for the numbers from 1 to 10, and to be able to reason in terms of visualized spatial patterns without counting-by-ones.
▶ Thus a child might reason that 8 can be partitioned into 4 and 4, and so on.

6.4.4: Partitioning Flashed Patterns to 10

▶ Flash a pairs-8 pattern. *How many dots did you see? Can you describe it another way?*
▶ Similarly using pairs patterns for 7, 9 and 10.

Purpose, Teaching and Children's Responses

▶ As for 6.4.2 and 6.4.3 above.

6.4.5: Combining Patterns Using 4-grids, 6-grids, 8-grids, 10-grids

▶   Flash an empty 4-grid. *How many squares did you see?* Flash a 4-grid with a row of two dots. *How many squares altogether? How many dots? How many empty squares?*
▶   Similarly using a 4-grid with one dot; and a 4-grid with three dots.
▶   Flash an empty 6-grid. *How many squares?* Flash a 6-grid with two rows of 2 dots. *How many squares altogether? How many dots? How many empty squares?*
▶   Similarly using other 6-grids, 8-grids and 10-grids.

Purpose, Teaching and Children's Responses

▶   An important goal is for children to visualize the grid patterns to answer questions such as how many dots or how many empty squares.
▶   As above, some children might attempt to count from one while visualizing patterns and other children might ascribe number to a group (for example, 2, 3 or 4 dots) without counting from one.
▶   Children should be encouraged to combine and partition the patterns without counting from one.

VOCABULARY

Squares. See also Key Topic 5.3.

MATERIALS

Spatial pattern cards for domino patterns from 1 to 6.
Spatial pattern cards for pairs patterns from 1 to 10.
4-grids, 6-grids, 8-grids, 10-grids showing various combinations.

ACKNOWLEDGMENT

These activities were adapted from the work of Paul Cobb and colleagues, and were further developed in the Mathematics Recovery project.

Key Topic 6.5: Finger Patterns

Purpose: To further develop facility with finger patterns for numbers in the range to 10.

LINKS TO LFIN

Main links: Part C, Finger Patterns.
Other links: Part C, Combining and Partitioning.
FNWS and BNWS, Levels 1–3; SEAL, Stage 1–2.

TEACHING PROCEDURES

6.5.1: Five Plus Patterns for 6 to 10

▶   Put hands out in front (that is, fingers seen). *Put your hands out in front of you like I am doing.* Raise five fingers. *Make five on one hand like I am doing.* Raise one finger on the other hand. *Now make one on the other hand. Look at your fingers. How many fingers raised altogether?*

▶   Similarly 5 and 2, ... 5 and 5.

▶   Put hands on head (that is, fingers not seen). *Put your hands on your head like I am doing.* Raise five fingers. *Make five on one hand like I am doing.* Raise one finger on the other hand. *Now make one on the other hand. Don't look at your fingers! How many fingers raised altogether? Check by looking at your fingers.*

▶   Similarly 5 and 2, ... 5 and 5.

▶   Repeat with children following instructions and not looking at the teacher's hands.

Finger patterns

Purpose, Teaching and Children's Responses

▶   An important goal of the activities in this key topic is for children to develop more sophisticated ways of reasoning about their finger patterns.

▶   Teaching should focus on creating situations in which children can explicitly reflect on their mathematical activity.

▶   Reflection of this kind can lead to more sophisticated ways of reasoning.

▶   To make a finger pattern for 5 some children might raise their fingers sequentially and others might raise them simultaneously.

▶   Teacher modeling, practice and reflecting on one's activity should be used to advance children from sequentially raising fingers to simultaneously raising fingers.

▶   Some children might count their fingers from one, and others count-on from five.

▶   As before, modeling and practice should be used to encourage children to count-on from five.

6.5.2: Partitioning Numbers 3 to 10

▶ *Put your hands on your head. Make three on your fingers. Can you make three another way? Check to see if you are correct. Look at each hand in turn and say the number that you have on that hand. Can you find another way to make three? Check. Say each of the numbers. How many ways can you make three?*
▶ *Make four on your fingers. Check. Find different ways to make four. How many ways?*
▶ Similarly for 5, 6, … 10.

Purpose, Teaching and Children's Responses

▶ Children might raise fingers simultaneously on one hand, and then raise fingers sequentially on the second hand in coordination with counting-on.
▶ Some children might systematically generate several partitions for a number, for example, 6 as 5 and 1, 4 and 2, 3 and 3, and so on. Systematic generation might involve, for example, putting down one more finger on one hand and raising one more on the other, for each new partition.

6.5.3: Doubles Plus One

▶ *Put your hands on your head. Show me 2 and 2 on your fingers. What does 2 and 2 make? Put up one more finger on one hand. What do you have now? What does 2 and 3 make?*
▶ *Now show me 3 and 3. What does 3 and 3 make? Put up one more finger. What do you have now? What does 3 and 4 make?*
▶ Similarly 4 and 4, and 4 and 5.

Purpose, Teaching and Children's Responses

▶ Children might count from one or count-on to work out a doubles plus one.
▶ Teacher questioning and child reflection can lead children to realize that the answer is always one more than the corresponding double.

6.5.4: Partitioning 10 Fingers

▶ *Put your hands out in front. Show me 10 on your fingers. Put down one finger. How many does that leave? What does 9 and 1 make? Show me that on your fingers.*
▶ Similarly, 8 and 2, 7 and 3, and so on.

Purpose, Teaching and Children's Responses

▶ Children might count from one, or count-on from five to figure out how many fingers remaining.
▶ Alternatively, they might reason that the number of fingers left is reduced by one each time another finger is put down.

VOCABULARY

Another way, 2 and 2, 3 and 3, and so on. See Key Topic 5.5.

MATERIALS

None.

ACKNOWLEDGMENT

These activities were adapted from the work of Paul Cobb and colleagues, and were further developed in the Mathematics Recovery project.

Key Topic 6.6: Equal Groups and Sharing

Purpose: To develop initial ideas of equal groups and equal sharing.

LINKS TO LFIN

Main links: Part D, Early Multiplication and Division.
Other links: FNWS and BNWS, Levels 1–4; SEAL, Stage 1.

TEACHING PROCEDURES

6.6.1: Describing Equal Groups

▶   Place out four 2-counter cards. *Here are some counters. What can you tell me about them? How many counters are there on each card? How many cards are there?*
▶   Place out three 4-counter cards. *Here are some counters. What can you tell me about them? How many counters are there on each card? How many cards are there?*
▶   Similarly 5 lots of 3, 6 lots of 2, 2 lots of 5, and so on.

Purpose, Teaching and Children's Responses

▶   The settings used in this key topic involve repeated equal groups.
▶   This key topic focusses on developing children's initial ideas, and it is important to observe carefully children's actions, language and ways of reasoning.
▶   As well, it is important to ensure that children have sufficient time to think about the tasks and to reflect on their own words and actions.
▶   Materials such as Unifix cubes are useful for settings involving equal groups. In each group the items can be arranged according to a standard pattern (for example, pairs patterns, die patterns).
▶   An important goal is for children to realize that each group has the same number of items.
▶   In the case of larger group sizes (for example, groups of 5) children might need to count the number of items from one to figure out how many in all. This is less likely to happen if a standard pattern is used, and the child is familiar with the pattern.

6.6.2: Organizing Equal Groups

▶ Place out a collection of 10 counters, with two of each of five colours. *Here are some counters. What can you see? Can you make a pattern with the counters? Tell me about the pattern.*
▶ Similarly 6 lots of 3, 4 lots of 5, 3 lots of 4, and so on.

Purpose, Teaching and Children's Responses

▶ An important goal is for children to arrange the counters into groups containing counters of the same color.
▶ After the child has arranged the counters into groups the goal is for children to realize that each group contains two (similarly three, four, and so on) counters.

6.6.3: Making Equal Groups

▶ Place out 8 plastic horses. *Here are some horses. Farmer Joe wants the horses put into twos. Can you put the horses into twos?*
▶ Place out 9 puppies. *Here are some puppies. Mary wants the puppies put into threes. Can you put the puppies into threes?*
▶ Place out 12 plastic horses. *Here are some horses. The farmer wants the horses put into fours. Can you put the horses into fours?*
▶ Similarly 10 into 5s, 8 into 4s, 12 into 3s, 14 into 2s, and so on.

Purpose, Teaching and Children's Responses

▶ Use identical plastic items for the groups (that is, same color, shape and size).
▶ Children might count from one (for example, 'one, two') when making each group.
▶ Alternatively they might say the number in each group without counting from one.

6.6.4: Describing Equal Shares

▶ Place out 2 dolls, each with 4 flowers. *Here are some dolls with flowers. How many dolls do you see? What do you notice about the number of flowers each doll has?*
▶ Place out 6 plates, each with 3 cookies. *Here are some plates of cookies. How many plates do you see? What do you notice about the number of cookies on each plate?*
▶ Similarly 5 lots of 2, 7 lots of 3, 4 lots of 5, and so on.

Purpose, Teaching and Children's Responses

▶ Attend carefully to the child's words and actions, in order to determine the strategy used.
▶ Children might count the number in each group from one.
▶ Alternatively, without counting from one, they might say the number in each group or otherwise indicate that each group has the same number.
▶ The teacher can model assigning the appropriate number to each equal group.

6.6.5: Organizing Equal Shares

▶ Place out a mixed collection of 5 red counters and 5 green counters. *Here are some counters. What can you see? The red counters are for Peter and the green counters are for Billie. Can you sort out the counters? What do you notice?*
▶ Place out a mixed collection of 5 white cows, 5 black cows and 5 brown cows, and three farmers. *Here are some cows. What can you see? Farmer Mary owns the white cows, Farmer Joe owns the black cows and Farmer Billie owns the brown cows. Can you sort out their cows? What do you notice?*
▶ Similarly 2 lots of 3, 3 lots of 6, 2 lots of 10, and so on.

Purpose, Teaching and Children's Responses

▶ Children might count each share from one or otherwise indicate that the shares are equal.

6.6.6: Partitioning into Equal Shares

▶ Place out 2 dolls and 10 plastic oranges. *Johnny and Kim have 10 oranges to share. Can you share out the oranges for them?*
▶ Place out 3 dolls and 12 puppies. *These three children are sharing 12 puppies. Can you share out the puppies for them?*
▶ Similarly 14 between 2, 9 among 3, 20 among 4.

Purpose, Teaching and Children's Responses

▶ Children might distribute one item to each group in turn.
▶ Alternatively they might use a less systematic sharing strategy, for example, distributing more than one item at a time, to one or more groups.

VOCABULARY

Twos, threes, fours, and so on, number each has, share, sharing, equal groups.

MATERIALS

Cards showing equal groups of twos, threes, fours, and fives.
Red/green counters.
Dolls, plates, and so on for equal shares.
Small plastic objects for sharing.

ACKNOWLEDGMENT

These activities were developed in the Mathematics Recovery project.

EXAMPLES OF WHOLE-CLASS LESSONS DESIGNED FOR THE CHILD AT THE PERCEPTUAL STAGE

Lesson 6(a)   **Hidden additions lesson**
Lesson 6(b)   **Spatial patterns**
Lesson 6(c)   **When the music stops**
Lesson 6(d)   **Linking double patterns with hidden addition and subtraction**

LESSON 6(A)

Title:	**Hidden additions**
Purpose:	To assist children in using more sophisticated strategies to solve hidden addition tasks.
Links to Key Topics:	6.3
Materials:	Insects (or something similar) and insect house for the overhead projector; counters for the overhead projector; Insect Addition game
Introductory Activities:	▶ Forward number words from different starting points. *Start counting from five. Start counting from twelve.*
	▶ Row tasks on the overhead projector.

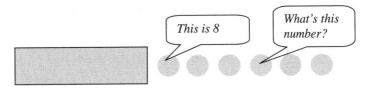

Main Focus:	The insect house.
	▶ The teacher places the insect house on the overhead projector.

▶ Five insects are hidden in the Insect house.

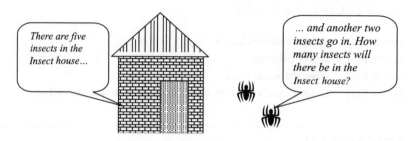

There are five insects in the Insect house...

... and another two insects go in. How many insects will there be in the Insect house?

▶ The lesson proceeds with further hidden addition tasks.
  – If children have difficulty solving hidden addition tasks use only one insect waiting to go into the insect house.
  – For those children still having difficulty, the teacher uses a transparent insect house. The insects are screened as soon as possible.

Group/Individual Tasks: Insect Game.

Pairs of children.

▶ A red spinner has the numerals 4 to 9 and a green spinner has the numerals 1, 2 and 3. The teacher can modify these numerals as needed.

▶ A player
  – spins the red spinner and puts this number of insects into the cup;
  – spins the green spinner and places this number of insects into the cup (or beside the cup);
  – works out how many insects there are in the cup.

▶ The other player checks the result.

▶ If correct, the player moves his/her counter this number of steps around the playing board (see below).

Conclusion/Summary: With the class together, the teacher gives further hidden addition tasks noting those students who are at figurative or counting-on stages and those who are still at the perceptual stage.

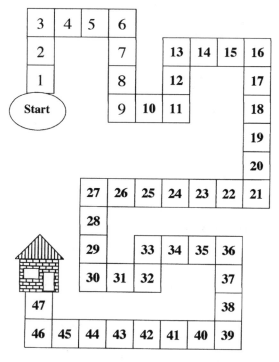

PLAYING BOARD

LESSON 6(B)

Title:	**Spatial patterns**
Purpose:	To introduce equal groups.
Links to Key Topics:	6.2, 6.4, 6.6
Materials:	Counters.
Introductory Activities:	▶ Flashing regular dot patterns.
	▶ Briefly display patterns such as this. *How many counters are there? Show your answer by holding up the correct number of fingers.*

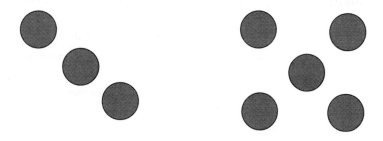

▶ Flashing irregular dot patterns.

▶ Briefly display patterns such as this. *How many counters are there? Show your answer by holding up the correct number of fingers using two hands.*

▶ Flashing equal groups.

Briefly display patterns such as this. *Tell us what you see.*

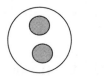

▶ Making equal groups

Get 8 counters. Make two equal groups.

Get 12 counters. Make three equal groups.

Group/Individual Tasks: Dot and numeral dominoes.

Pairs of children play dominoes using domino cards like these:

Conclusion/Summary: – With the class together, the teacher flashes some dot patterns and asks children to say what they are.

– The teacher may decide to give a verbal question such as this:

▶ *There are three plates. Each plate has two dots on it. How many dots are there altogether?*

▶ *There are eight apples. If we share them equally between four people how many will they each get?*

LESSON 6(C)

Title: **When the music stops [or when the whistle blows]**

Purpose: To assist children in forming equal groups and to focus on the number of groups, the number of items in each group, and how many items altogether.

Links to Key Topics: 6.4, 6.6

Materials: Large cardboard numeral, counters, music or whistle (or some other attention-seeking device).

Introductory Activities:
▶ Teacher asks the children (individually or in pairs) to make three groups with four in each group.

▶ Equal groups are flashed and the teacher asks questions like: *How many groups? How many in each group?*

Main Focus:
▶ This is a beginning activity introducing children to equal groups.
  − The teacher plays the music. Children skip around in a set area (inside or outside).
  − The teacher stops the music and holds up a large numeral.
  − The children form groups with the number of children in each shown by the numeral held up by the teacher.
  − *How many groups? How many in each group?*
  − Plastic hoops could be placed on the ground in which children form their groups. This may further emphasize the notion of a group.

Group/Individual Tasks: Equal Groups Game.

  − Equipment: red die (or spinner) with numerals 2, 3, 4; green die (or spinner) with numerals 1, 2, 3, 4; lids or similar containers; counters or similar items. Players take turns to:

▶ roll red die − takes this number of lids;

▶ roll green die − puts this number of items in each lid;

▶ work out how many items there are altogether. (This can be marked on a recording sheet if required.)

Conclusion/Summary: With the whole class the teacher flashes some equal groups. *How many groups? How many in each group? How many altogether?*

▶ The teacher covers groups each containing two items and asks if anyone can find out how many items there are altogether.

▶ *There are two Smileys in each house. How many are there altogether?*

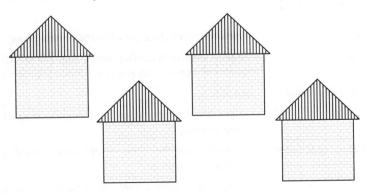

LESSON 6(D)

Title: **Linking doubles patterns with hidden additions and subtractions**

Purpose: To assist children in using doubles patterns to solve addition tasks.

Links to Key Topics: 6.3, 6.4, 6.5

Materials: Sets of doubles cards and corresponding numeral cards for Fish (see diagram below).

Introductory Activities:
▶ Teacher: *Hold up two fingers on each hand and put them on your head like rabbit's ears. How many fingers are you holding up?*

▶ Repeat this task for other doubles.

▶ Flash doubles patterns (for example, 3 red and 3 blue counters). *How many dots are there? How did they look?*

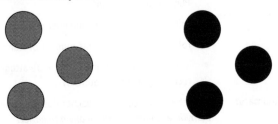

Briefly display then screen three counters. Place three more counters next to the screen. *How many counters are there altogether?*

Main Focus:
▶ Briefly display then screen six counters. Take three counters out from under the screen. *There are six counters. If I take out three counters, how many are left under the screen?*

▶ Give similar tasks. Note how children are solving these tasks.

▶ Are they using their doubles knowledge?

▶ Can they link addition and subtraction – are they using the previous addition to help solve the subtraction task?

▶ Are there links between the settings in the introductory segment of this lesson (rabbit's ears, doubles patterns and the hidden tasks)?

Group/Individual Tasks: Doubles Fish.

- The game of Fish is played using doubles cards and numeral cards.

- Children make pairs of cards by matching a doubles card with a numeral card. Initially regular dice patterns can be used and different patterns can gradually be introduced.

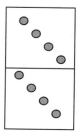 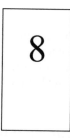 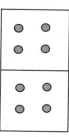

Conclusion/Summary: The whole class is assembled.

- The teacher flashes some doubles patterns (using different configurations). *How many dots are there?*

Some hidden addition and subtraction tasks are given using doubles.

7

Teaching the Figurative Child

This chapter focusses on teaching children at the Figurative Stage on the SEAL (Stages of Early Arithmetical Learning – see Chapter 1). First, we provide a thorough description of the knowledge and strategies typical of this stage. This is followed by detailed descriptions of six key topics which can provide a basis for teaching the child at the Figurative Stage. The six key topics consist of a total of 44 teaching procedures.

THE TYPICAL FIGURATIVE CHILD

This section provides an overview of a typical child at the Figurative Stage, that is at Stage 2 of the Stages of Early Arithmetical Learning. Table 7.1 sets out the stage and indicative levels for the figurative child on the models pertaining to FNWSs, BNWSs, Numeral Identification and Tens and Ones. The overview discusses the first four aspects of early number knowledge listed in Table 7.1, and also discusses knowledge of spatial patterns and facility with finger patterns.

Table 7.1 Stage and levels of a typical figurative child

Model	Stage/level
Stage of Early Arithmetical Learning (SEAL)	2
Level of Forward Number Word Sequences (FNWSs)	4
Level of Backward Number Word Sequences (BNWSs)	3
Level of Numeral Identification	2
Level of Tens and Ones Knowledge	0

Stage of Early Arithmetical Learning

The child at the Figurative Stage can solve additive tasks involving two screened collections and, when doing so, counts from one. Thus on an additive task involving, say, a collection of 6 red counters and a collection of 3 blue counters, both of which are screened, the child might count from one to six to count the counters in the first collection, and then continue counting from seven to nine to count the counters in the second collection. From a cognitive perspective it seems necessary for the child to count from one in order to give meaning to six, that is, the number of counters in the first collection. The child typically cannot make sense of subtractive tasks such as missing addend, missing subtrahend or comparison tasks.

FNWSs, BNWSs and Numeral Identification

Children at this stage typically are facile with FNWSs in the range 1 to 30, but might not know the FNWS to 100. They might not be able to say the number word after some of the numbers beyond 30, for example, 49 or 80. Typically these children are facile with BNWSs to 10, but might have difficulty in producing the BNWS from 23 to 16, and from 15 to 10. In similar vein they might have difficulty saying the word before number words such as 12, 15, 20, 21 and 30, or might use a dropping back strategy to work out the number word before a given number word. Some children at this stage say an incorrect decade number when saying BNWSs, for example, '52, 51, 40, 49, 48'. Children at this stage typically can recognize and identify 1-digit numerals and numerals in the teens, although the numeral '12' might be incorrectly identified as 'twenty' or 'twenty-one'. Common also at this stage is the error of 'digit reversal' when identifying 2-digit numbers. Thus for example '27' is named as 'seventy-two'.

Spatial Patterns and Finger Patterns

Children at this stage typically will have facile finger patterns for numbers in the range 1 to 5 and some children will also have finger patterns for numbers in the range 6 to 10. Children may also have sound knowledge relating to spatial patterns (for example, pair-wise patterns for the numbers from 1 to 10) and are able to subitize random arrays of up to 5 items. They can regard a pattern for 6 for example, alternatively as 3 twos or 2 threes, and a pattern for 8, as 2 fours or as 5 and 3.

The Way Forward

The figurative child has a relatively extensive body of knowledge relating to number words and numerals and early counting. Knowledge of FNWSs can be extended to 100 and beyond, and knowledge of BNWSs can be extended to 30 and beyond, and children can learn to count forwards and backwards by tens to at least 100, and similarly by fives forward and backward. Counting by twos can also be learned. Knowledge of numerals and numeral sequences can be extended to 100 using settings such as the numeral roll and the hundred square as well as the numeral track. Children can be presented with additive and subtractive tasks involving screened collections in ways conducive to the development of counting-on and counting-back. In similar vein, additive and subtractive tasks involving rows of items, some of which are screened, can support the development of these strategies. The figurative child can learn to combine and partition numbers in the range 1 to 10 in settings emphasizing spatial patterns and can learn to structure numbers in the range 1 to 10 in ways which involve building to five and ten, or building onto five. Finally, instruction can focus on developing initial multiplicative and divisional knowledge through tasks involving equal groups and sharing (for example, tasks involving equal groups which are screened).

The next section contains six key topics and 44 teaching procedures, which form the basis of an appropriate instructional programme for the figurative child.

Key Topic 7.1: Number Word Sequences from 1 to 100

Purpose: To develop knowledge of number word sequences in the range 1 to 100.

LINKS TO LFIN

Main links: FNWS and BNWS, Levels 1–5.
Other links: SEAL, Stage 3.

TEACHING PROCEDURES

7.1.1: Saying Short FNWSs

▶   *Start from 28 and count up to 34. Now start from 44 and count up to 52.*
▶   *Similarly, 65 to 73, 88 to 96, and so on.*

Purpose, Teaching and Children's Responses

▶   The activities in this key topic focus on number word sequences alone, that is, not in conjunction with numerals or collections or rows of items (see the first Note at the end of Teaching Procedure 6.1 for a more extensive explanation of this point). Again, we believe it is important to include these activities as well as the activities in Key Topic 7.2, which focus on numerals in conjunction with number words, and similarly it is important to include the activities in Key Topic 7.3 which focuses on counting-on and counting- back.
▶   Children might make an error at the decade number word, for example 'forty-nine, sixty, sixty-one, and so on.'

7.1.2: Saying Short BNWSs

▶   *Start from 33 and count back to 26. Now start from 48 and count back to 35.*
▶   *Similarly, 52 back to 47, 85 back to 77, 94 back to 86, and so on.*

Purpose, Teaching and Children's Responses

▶   There are two relatively common errors with these tasks. The first is to omit the decade number (for example, 'forty-one, thirty-nine, thirty-eight, and so on'). The second is to say the decade number, which is ten less than the correct decade number (for example, 'forty-one, thirty, thirty-nine, thirty-eight, and so on').
▶   A likely explanation for the first error is that the child thinks of the number words in the forties, for example, as consisting of the two parts, forty and a ones number. From this reasoning forty-one seems to be the last number in the forties (going backwards), and the number after forty-one is the highest number in the thirties, that is, thirty-nine.
▶   A likely explanation for the second error is that in order to work out what comes before forty-one, for example, the child's only recourse is to think about the backward sequence of decade numbers. The child works out that thirty comes before forty and so says 'thirty'. The child then continues by saying the number words in the thirties backwards (that is, 'thirty-nine, thirty-eight, … and so on).
▶   These errors can be addressed by teaching procedures based on settings in which the numerals are in sequence (see Key Topic 6.2 for procedures involving numeral sequences, numeral track and numeral roll), because in these procedures the children have an opportunity to see the numerals in sequence (that is, 42, 41, 40, 39, 38, …), as well as to hear and say the corresponding number words.

7.1.3: Saying One, Two or Three Numbers After a Given Number

▶ *I'm going to say a number and I want you to say the next number after the number I say. What comes after 7, after 12, after 19, after 50, after 77, and so on.*
▶ *This time say the next two numbers after the number I say. What are the next two numbers after 8, after 35, and so on.*
▶ *This time say the next three numbers after the number I say. What are the next three numbers after 11, after 49, and so on.*

Purpose, Teaching and Children's Responses

▶ As explained in Key Topic 6.1, facility with saying one, two or three number words forwards forms an important basis for counting on strategies (counting-up-from and counting-up-to).
▶ Children might use fingers to keep track of two or three number words.
▶ Children typically have more difficulty when answers bridge a decade (for example, three numbers after 68).

7.1.4: Saying One, Two or Three Numbers Before a Given Number

▶ *I'm going to say a number and I want you to say the number before the number I say. What comes before 10, before 21, before 28, before 40, before 66, before 80, before 92, and so on.*
▶ *This time say the next two numbers before the number I say. What are the two numbers before 6, before 31, and so on.*
▶ *This time say the next three numbers before the number I say. What are the three numbers before 16, before 41, and so on.*

Purpose, Teaching and Children's Responses

▶ As explained in Key Topic 6.1, facility with saying one, two or three number words backwards forms an important basis for counting down strategies (counting-down-from and counting-down-to).
▶ As for 7.1.2 above, children might use fingers to keep track, and typically have more difficulty when answers bridge the decade.
▶ Children might say a forward number word sequence to generate the number before a given number.

7.1.5: Counting the Number of Jumps Forwards from a to b

▶ *I'm going to count the number of jumps from one number to another. How many jumps from 6 to 8? Six –, seven, eight – two jumps. Now you count the number of jumps from 4 to 6.*
▶ *Similarly, from 10 to 12, from 16 to 19, from 22 to 25, 59 to 63, 87 to 92, and so on.*

Purpose, Teaching and Children's Responses

▶ Children might use fingers to keep track.
▶ Children might count each number after the starting number (for example, from 6 to 8, counting 7 as one and 8 as two).
▶ Alternatively, children might count each jump (for example, from 6 to 7 is one jump and so on).

▶   In the case of counting forwards, it is not easy to distinguish between these two strategies because from the observer's viewpoint the strategies appear to be the same.

▶   In the case of the first mentioned strategy, children might make the error of counting the initial number as one (for example, from 6 to 8, counting 6 as one, 7 as two and 8 as three).

▶   These errors can be addressed by teaching procedures based on settings involving numeral sequences (this includes the numeral track and numeral roll. See Key Topic 7.2) and rows of dots (see Key Topic 7.3).

7.1.6: Counting the Number of Jumps Backwards from b to a

▶   *I'm going to count the number of jumps from one number back to another. How many jumps from 10 back to 7? Ten –, nine, eight, seven, – three jumps. Now you count the number of jumps from 12 back to 10.*

▶   *Similarly from 18 back to 15, from 30 back to 26, 63 back to 58, 71 back to 67, 95 back to 92.*

Purpose, Teaching and Children's Responses

▶   Children might use fingers to keep track.

▶   Children might count each number commencing from the starting number (for example, from 10 to 7, counting 10 as one, 9 as two, and 8 as three).

▶   Alternatively, children might count each jump (for example, from 10 to 9 is one jump and so on).

▶   In the case of counting backwards the observer can distinguish between these two strategies. The first involves counting (saying) the starting number and the second does not involve counting (or saying) the starting number.

▶   In the case of the first mentioned strategy, children might make the error of counting the final number (for example, from 10 to 7, counting 10 as one, 9 as two, 8 as three, and 7 as four).

▶   As above (see 7.1.5), these errors can be addressed using procedures based on settings involving numeral sequences and rows of dots.

7.1.7: Forwards and Backwards Using the Sequence of Decade Numbers from 10 to 100

▶   *I'm going to count by tens to 100. Ready, 10, 20, … 100. Now you count by tens.*

▶   *This time I'm going to count backwards by tens from 100. Ready, 100, 90, … 10. Now you count backwards by tens.*

▶   *This time start from 50 and count forwards by tens. From 80 and so on.*

▶   *Now start from 60 and count backwards by tens. From 40 and so on.*

▶   *This time start from 30 and count forwards 3 tens. From 60 and count forwards 3 tens and so on.*

▶   *This time start from 70 and count backwards 1 ten. From 50 and count backwards 2 tens and so on.*

Purpose, Teaching and Children's Responses

▶   The purpose of this procedure is to begin to extend to the sequence of numbers by tens (that is, the decade numbers), some of the tasks in the above procedures (for example, going forwards or backwards 1, 2 or 3 counts) which are based on the sequence by ones (that is, one, two, three, and so on).

▶   Children might say the teen numbers instead of the corresponding decade numbers (for example, ten, twenty, thirteen, fourteen, and so on).

▶   When saying the sequence backwards children might say 'twelve' instead of 'twenty' (for example, '… forty, thirty, twelve' or 'fourteen, thirteen, twelve').

VOCABULARY

One, two, three, … one hundred, count by tens, count backwards by tens. See also Key Topic 6.1.

MATERIALS

None.

ACKNOWLEDGMENT

These activities were adapted from the work of Robert Wright, and were further developed in the Mathematics Recovery project.

Key Topic 7.2: Numerals from 1 to 100

Purpose: To develop knowledge of numerals in the range 1 to 100.

LINKS TO LFIN

Main links: Numeral Identification, Levels 1–3.
Other links: FNWS and BNWS, Levels 1–5; SEAL, Stages 1–3.

TEACHING PROCEDURES

7.2.1: Sequencing Numerals

▶ Place out the cards from 26 to 30, randomly arranged. *Put these cards in order.* Direct the child to arrange cards left to right in increasing order. *Now say the numbers as you point to them.*
▶ Similarly order the cards from 41 to 45, 56 to 65, 88 to 96, and so on.

Purpose, Teaching and Children's Responses

▶ Activities in which children have to arrange numerals in correct sequence are useful as well as activities with settings based on sequences of numerals (for example, numeral sequences, numeral tracks, numeral rolls – see below).
▶ Observing children's strategies for sequencing numerals can provide interesting insights into children's knowledge and ways of thinking.
▶ Sequences with fewer numerals are easier for children to sequence.
▶ Sequences within a decade are easier than sequences that bridge two decades.
▶ It is not unusual for some children to sequence numerals according to the right-hand digit only (that is, the 'ones' digit). Thus the numerals from 56 to 65 for example, might be sequenced thus: 61, 62, 63, 64, 65, 56, 57, and so on. Children might read the numerals in the order just shown and not seem to be aware that the numerals are out of order.

7.2.2: Sequencing Decade Numerals

▶   Place out the decade cards from 10 to 40, randomly arranged. *Put these cards in order.* Direct the child to arrange the cards left to right in increasing order. *Now say the numbers as you point to them.*
▶   Similarly order the cards from 50 to 80, 20 to 70, 10 to 100, and so on.

Purpose, Teaching and Children's Responses

▶   Sequences with fewer numerals are easier for children to sequence.
▶   Observe closely to see if children confuse decade names (for example, thirty, forty) with teen names (for example, thirteen, fourteen).
▶   Some children have a preference for arranging the numeral cards vertically rather than horizontally.

7.2.3: Sequencing Off-Decade Numerals

▶   Place out the following cards, 6, 16, 26, 36, randomly arranged. *Put these cards in order.* Direct the child to arrange the cards left to right in increasing order *Now say the numbers as you point to them.*
▶   Similarly order the cards from 34, 44, ... 64; 49, 59, ... 99; 5, 15, 25, ... 95; 2, 12, 22, ... 92; and so on.

Purpose, Teaching and Children's Responses

▶   These activities can form a basis for incrementing and decrementing by tens (for example, Key Topic 8.3).
▶   Sequences with fewer numerals are easier for children to sequence.
▶   The following complementary task can also be used. Using a hundred square (see below) have the child put a counter on each of 2, 12, 22 ... 92.

7.2.4: Ordering 2-Digit Numerals

▶   Place out four cards in the range 1 to 30 (for example, 12, 20, 18, 25), randomly arranged. *Put these cards in order.* Direct the child to arrange the cards left to right in increasing order. *Now say the numbers as you point to them.*
▶   Similarly four or more cards in the range 1 to 40, 40 to 100, 1 to 100, and so on.

Purpose, Teaching and Children's Responses

▶   The term 'ordering' rather than 'sequencing' is used in cases where the numerals to be ordered do not follow a regular sequence.
▶   Observing children's strategies for ordering numerals can provide interesting insights into children's knowledge and ways of thinking.
▶   Children typically will identify numerals before ordering them. Thus a common strategy is for children to figure out the names of the numerals and use the name as a basis for ordering the numerals.
▶   Particularly in the case of ordering (cf. sequencing) numerals, children might need to think hard for some time to read (that is, name/identify) the numerals.
▶   Errors in ordering numerals can result from incorrect identification of the numeral (for example, 12 is read as 'twenty-one' and is placed after 18 and 20).

Ordering 2-digit numerals

▶ Sets with fewer numerals are easier for children to order. Particularly when the children's strategy involves reading the names of the numerals, they might need to identify a numeral several times during their attempts to order the numerals.
▶ Some children might be able to reason that if *a* comes before *b* and *b* comes before *c*, then *a* comes before *c* (the principle of transitivity).
▶ The teacher could show the child where the numerals in question appear on a numeral roll or a hundred square.

7.2.5: Numeral Recognition

▶ Place out ten cards in the range 1 to 30, randomly arranged. *Point to the number 12. Point to the number 20. Point to the number 25.*
▶ Similarly using cards in the range 1 to 50, 50 to 100, 1 to 100.

Purpose, Teaching and Children's Responses

▶ For some children recognition of particular numerals is easier than identification of those numerals.
▶ Children can and should learn the names of 2-digit numerals at a time when they might have little understanding of the place value associated with the numerals (as noted in Key Topic 6.2).

7.2.6: Numeral Identification

▶ Place out the cards from 30 to 40, randomly arranged. Point to the card for 34. *What number is this?* Point to the card for 30. *What number is this?*
▶ Similarly using the cards from 56 to 65, 80 to 100, and so on, and 1 to 100.

Purpose, Teaching and Children's Responses

▶ Children might make the error of 'digit-reversal', for example, the numeral 74 is named as 'forty-seven'. This can be attributed to inappropriate application to numerals beyond 20 of a process for identifying some of the teen numerals by first reading the right-hand digit. In this process the numeral 16 for example, is read by first looking at the '6' and saying 'six', and then saying 'teen'. Thus the child focusses on the right-hand digit first to determine part of the number name, rather than focussing on the whole numeral.

7.2.7: Numeral Tracks in Range 20 to 100

▶ Place out the numeral track from 36 to 45, with numerals uncovered. *Watch me as I count forwards and backwards.* Point to each numeral in turn, while counting from 36 to 45, and then from 45 to 36. *Now you count forwards and backwards and point to each number in turn.*
▶ Close the lids on the numeral track and repeat the previous activity, uncovering each numeral after saying the corresponding number word.
▶ Place out the numeral track from 36 to 45, with all numerals covered. Uncover the numeral 38. Place a green cube adjacent to the numeral 38 to mark it (or place a transparent counter on the numeral). *What number is that?* Leave the numeral 38 uncovered, and place a red cube adjacent to the lid covering the numeral 41. Point to the lid covering 41. *What number is that?* Direct the child to uncover the numeral 41. *Were you correct?*
▶ Similarly for other pairs of numerals from 36 to 45.
▶ Uncover the numeral 40. Place a green cube adjacent to the numeral 40 to mark it. *What number is that?* Leave the numeral 40 uncovered, and place a red cube adjacent to the lid covering the numeral 37. Point to the lid covering 37. *What number is that?* Direct the child to uncover the numeral 37. *Were you correct?*
▶ Similarly for other pairs of numerals from 36 to 45.
▶ Repeat the previous activities using the numeral tracks from 36 to 65 and so on.

Purpose, Teaching and Children's Responses

▶ Children have more difficulties with tasks which bridge two decades, particularly in cases which involve moving backwards on the numeral track.
▶ Numeral tracks which bridge two decades (for example, the numeral track from 36 to 45) can be useful in addressing children's difficulties associated with the start or end of a decade.
▶ Numeral tracks containing 20, 30 or more numerals can be made simply by putting tracks end to end or one above another (for example, 36 to 45 is placed above 46 to 55 and 56 to 65).
▶ Translucent paper can be used to highlight for children points at which they are having difficulty, for example, a sheet of red translucent paper is placed around the numerals 39 and 40.

7.2.8: Numeral Roll 1 to 100

▶ Display the numeral roll from 1 to 100. *This numeral roll shows the numbers from 1 to 100 in order. You say the numbers with me that we can see as I roll it. Ready, 38, 39, 40, 41, ... 60. Now let's say the numbers backwards. Ready, 60, 59, ... 38.*

Numeral roll 1 to 100

Purpose, Teaching and Children's Responses

▶ As stated in Key Topic 6.2, we believe numeral rolls can help to develop children's knowledge in distinctive ways.

▶ Concerning the previous point, for children the order and sequence of numerals is likely to be more apparent in the case of the numeral roll compared with settings such as the hundred square.

▶ Children will use their knowledge of number word sequences to read in sequence the numerals on the numeral roll.

▶ A slotted card can be used with the numeral roll (see Key Topic 6.2).

▶ Children can cut up a numeral roll (using scissors) into decades (1–10, 11–20, and so on) which can be rearranged to make a hundred square.

7.2.9: Hundred Square

▶ Place out a hundred square. *Look away while I cover a number.* Place a small cover over the numeral 32. *What number is under there? Take off the cover and see if you are correct.*

▶ Similarly with the numerals 45, 74, 88, and so on.

▶ *Look away while I cover some numbers.* Place a small cover (for example, an opaque counter) over each of the numerals 53, 54, 63, 64. Point to the 53 square. *What number is under there?* Remove the cover. *Were you correct?*

▶ Similarly for a five numeral cross such as 74, 75, 76, 65, 85, and a 3 × 3 square of numerals.

Purpose, Teaching and Children's Responses

▶ Children might confuse one more/less with ten more/less, for example the numeral immediately below 75 on the hundred square is called 76 instead of 85.

▶ Confusion of one more/less with ten more/less can be addressed by working more with settings such as the numeral track and the numeral roll, which highlight the sequence of the numerals.

Hundred square (1)

Hundred square (2)

7.2.10: Blank Hundred Square

▶ Write the 'fours' column on the square (that is, 4, 14, 24, and so on). *Read the numbers that I have written. What patterns do you see?* Place a marker on the 26 square. *What number goes here?* Write the numeral 26 in the appropriate square.

▶ Similarly 27, 28, 23, and so on.

▶ Similarly for other columns or rows on the square.

Purpose, Teaching and Children's Responses

▶   As for 7.2.9 above.
▶   Allow sufficient time for the child to think about the task.
▶   Try to determine the child's strategy, for example, which given number(s) do they start from. Do they count subvocally or aloud by ones or by tens, and so on.

VOCABULARY

Hundred square. See also Key Topic 6.2.

MATERIALS

Numeral cards for each number in the range 1 to 100.
Numeral tracks in the range 20 to 100.
Numeral roll from 1 to 100.
Hundred square.
Blank hundred square.
Small green cube and small red cube (or transparent counters) to use as markers.

ACKNOWLEDGMENT

Most of these activities were adapted from the work of Robert Wright, and were further developed in the Mathematics Recovery project. The numeral track was developed by Garry Bell (Southern Cross University). Adaptations to the use of the numeral track were developed by Ian Gray (New South Wales). The slotted card was developed by Sandy Norris and Dean Fowler (South Carolina).

Key Topic 7.3: Counting-On and Counting-Back

Purpose: To develop strategies involving counting-on and counting-back.

LINKS TO LFIN

Main links: SEAL, Stages 3–4.
Other links: FNWS and BNWS, Levels 1–5.

TEACHING PROCEDURES

7.3.1: Counting Items in Two Screened Collections

▶   Place out 14 red counters. *Here are 14 red counters. I'm going to cover those 14 counters.* Place out 3 green counters. *Here are 3 green counters. I'm going to cover the 3 green counters.* Place a screen over the 3 green counters. *How many counters are there in all?*
▶   Similarly, with 23 red and 4 green, 41 red and 2 green, and so on.

Purpose, Teaching and Children's Responses

▶ It is important to give children ample time to think and to reflect on their thinking.

▶ On the tasks in this key topic it is particularly important to observe and attempt to fully understand the child's strategy (see the last point in this list).

▶ See notes in Key Topic 6.3 about sizes of the numbers in the collections. For children at this level the first number can range in size to 100 and beyond, and the second number should range up to about 5 or 6. Thus the second number should be a number which the child can keep track of when counting-on.

▶ An important strategy for children to develop is to count-on from the first and larger collection, and keep track of the number of counters in the second, smaller collection.

▶ Children's strategies for keeping track include using fingers, explicit double counting and recognizing a temporal pattern of counts (for example, recognizing a pattern of three counts, that is, 'fifteen, sixteen, seventeen').

▶ This counting-on strategy is also called counting-on-from/counting-up-from.

▶ Some children solve some or all of these tasks using strategies which wholly, or in part, do not involve counting-by-ones (for example, 7 and 5 is 5 and 5 and 2 more). These kinds of strategies are called non-count-by-ones and are considered to be generally more advanced than strategies involving counting-by-ones only.

7.3.2: Counting Items in a Row with Some Items Screened

▶ Place out a row of 30 dots. Place a marker (for example, a small green cube) adjacent to the twenty-first dot (or a transparent counter on the dot). *This is number 21.* Place a small screen over the twenty-second, twenty-third and twenty- fourth dots. *There are three under here.* Point to the screen. Place a different colored marker (for example, a small red cube) on the twenty-fifth dot. *What number is this one?*

▶ Similarly, with eighteenth and twenty-first dots, twenty-fourth and twenty-eighth dot, and so on.

Purpose, Teaching and Children's Responses

▶ The notes above and in Key Topic 6.3 about sizes of numbers apply here as well. Thus the number of covered dots would range from 1 to 3 or 4.

▶ An important strategy for children to develop is to count-on and keep track of their counts (also called counting-on-from/counting-up-from).

▶ Children might answer one less than the correct answer (for example, 24 rather than 25) because they stop counting at the last covered counter.

▶ As above (see the last point in 7.3.1) children might use a non-count-by-ones strategy on some or all of these tasks.

7.3.3: Missing Addend Tasks

▶ Briefly display and then screen 7 red counters. *Here are 7 red counters.* Ask the child to look away while screening 3 green counters. *With the green counters there are 10 counters in all. How many green counters are there?*

▶ Similarly, with 10 red and 4 green, 15 red and 2 green, and so on.

Purpose, Teaching and Children's Responses

▶   The notes above and in Key Topic 6.3 about sizes of numbers apply here as well. Thus the number of the missing addend would range from 1 to no more than 5 or 6, and the known addend should not be smaller than the missing addend.

▶   It is common for children when first presented with a missing addend task to misinterpret it as involving addition (for example, 7 and 10 rather than 7 and what make 10).

▶   An important strategy for children to develop is to count-on and keep track of the number of counts (for example, 8 is one, 9 is two, 10 is three; or use fingers to keep track).

▶   This counting-on strategy is also called counting-on-to/counting-up-to.

▶   As above (see the last point in 7.3.1) children might use a non-count-by-ones strategy on some or all of these tasks (for example, I know 8 and 2 make 10, so 8 and 3 make 11).

7.3.4: Removed Items Tasks

▶   Briefly display and then screen 11 red counters. Ask the child to look away while removing 2 counters. Briefly display and then screen the 2 counters. *There were 11 counters and I removed 2. How many counters are left?*

▶   Similarly, with 7 remove 3, 15 remove 4, and so on.

Purpose, Teaching and Children's Responses

▶   In similar vein to earlier points, the number of removed items should be in the range of 1 to no more than 5 or 6, and should not be greater than the number of items remaining.

▶   An important strategy for children to develop is to count back and keep track of the number of counts (eleven, ten, nine).

▶   This is called counting-down-from/counting-back-from/counting-off from.

▶   As above (see the last point in 7.3.1) children might use a non-count-by-ones strategy on some or all of these tasks.

7.3.5: Missing Subtrahend Tasks

▶   Briefly display and then screen 8 red counters. *Here are 8 red counters.* Ask the child to look away while removing and screening 3 of the red counters. *There were 8 red counters and I removed some and now there are only 5. How many did I remove?*

▶   Similarly, with 14 to 11, 13 to 9, and so on.

Purpose, Teaching and Children's Responses

▶   As above, the number of the missing subtrahend should be in the range of 1 to no more than 5 or 6, and should not be greater than the number of items in the difference.

▶   These tasks typically are more difficult for children than missing addend or removed items tasks involving similar numbers.

▶   This task tends to invoke a strategy involving counting back from the minuend (that is, 8 in the above example) to the known difference (that is, 5 in the above example) and keeping track. This is called counting-down-to/counting-back-to.

▶   Some children might solve this task by counting up from the known difference to the minuend and keeping track. This strategy is called counting-on-to/counting-up-to.

▶ Some children might attempt to solve this task by counting the known difference off from the minuend, that is, using the strategy of counting-off-from. The difficulty with this strategy is that for many of these tasks, it involves keeping track of a relatively large number of backward counts, which is usually very difficult and error prone.

▶ As above (see the last point in 7.3.1) children might use a non-count-by-ones strategy on some or all of these tasks.

7.3.6: Subtractive Tasks Using a Row

▶ Place out a row of 20 dots. Place a marker adjacent to the ninth dot (or place a transparent counter on the dot). *This is number 9.* Place a small screen over the tenth, eleventh and twelfth dots. Place a marker on the thirteenth dot. *This is number 13.* Point to the screen. *How many are under here?*

▶ Similarly, with seventh and tenth dots, fourteenth and eighteenth dots, and so on.

▶ Place out a row of 20 dots. Place a marker adjacent to the tenth dot. *This is number 10.* Place a small screen over the ninth and eighth dots. *There are two under here.* Point to the screen. Place a different colored marker on the seventh dot. *What number is this one?*

▶ Similarly cover three before seventeenth dot and so on, and using the row of 30 dots.

▶ Place out a row of 20 dots. Place a marker adjacent to the fifteenth dot. *This is number 15.* Place a small screen over the fourteenth, thirteenth, and twelfth dots. Place a marker on the eleventh dot. *This is number 11.* Point to the screen. *How many are under here?*

▶ Similarly cover two before fourteenth dot and so on, and using the row of 30 dots.

Purpose, Teaching and Children's Responses

▶ This task tends to invoke a strategy involving counting back from the higher number to the lower number and keeping track. This is called counting-down-to/counting-back-to.

▶ Children might answer one less than the correct answer because they stop counting at the last covered counter.

7.3.7: Comparison Tasks

▶ Place out 6 green counters. *Here are 6 horses.* Place out 4 white cubes. *Here are 4 jockeys. If each jockey got onto a horse, how many horses would not have a jockey?*

▶ Similarly with 10 counters and 7 cubes, 13 counters and 8 cubes, and so on.

Purpose, Teaching and Children's Responses

▶ The number of the unknown difference should be in the range of 1 to no more than 5 or 6, and should not be greater than the number of items in the smaller collection.

▶ Children might solve these kinds of tasks by counting-up-to, counting-down-to or using a non-count-by-ones strategy.

▶ These tasks typically are more difficult for children than missing addend or removed items tasks involving similar numbers.

Comparison tasks

VOCABULARY

How many left, how many removed, counting-on, counting-back. See Key Topic 6.3.

MATERIALS

Counters of two colors.
Rows of 20 dots and 30 dots.
Small green cube and small red cube (or transparent counters) to use as markers.
Small screens.

ACKNOWLEDGMENT

Most of these activities were adapted from the work of Leslie Steffe and the work of Robert Wright.

Key Topic 7.4: Combining and Partitioning Involving Five and Ten

Purpose: To develop facility with using five and ten to combine and partition numbers in the range 1 to 10.

LINKS TO LFIN

Main links: Part C, Base-five.
Other links: Part C, Combining and Partitioning, Part C, Finger Patterns. SEAL, Stage 3, FNWS and
BNWS, Levels 1–3.

TEACHING PROCEDURES

7.4.1: Combining Numbers to 5

▷ Place out a five(3) frame. *How many dots do you see? How many empty squares do you see? How many dots and empty squares altogether?*
▷ Similarly with other five(n) frames.
▷ Flash a five(3) Frame. *How many dots did you see? How many empty squares did you see? How many dots and empty squares altogether?*
▷ Similarly with other five(n) frames.

Combining numbers to 5

Purpose, Teaching and Children's Responses

▷ An important goal is for children to develop their knowledge of the patterns on the five frame for the numbers from 1 to 5.
▷ Related to this, an important goal is for children to be able to say the number of dots or empty squares without counting from one.
▷ Children might reason that the number of dots and empty spaces altogether is always five.

7.4.2: Partitioning 5

▶ Place out an empty five frame, and place 3 red and 2 green counters in the squares. *How many counters altogether? How many red counters? How many green counters?*
▶ Similarly with other partitions of 5 (2 and 3, 4 and 1, 1 and 4).
▶ Flash a five frame with 4 red and 1 green counters in the squares. *How many counters altogether? How many red counters? How many green counters?*
▶ Similarly with other partitions of 5 (1 and 4, 3 and 2, 2 and 3).

Purpose, Teaching and Children's Responses

▶ An important goal is for children to visualize the partitions and to reason conceptually about the partitions. Thus the goal is for children to reason in terms of their visualized images of the partitions.
▶ For children who have continued difficulty, increase the time interval of flashing and then gradually decrease the time interval.

7.4.3: Combining 5 and a Number in the Range 1 to 5

▶ Place out an empty ten frame. Place 5 red counters in the upper row, and 2 green counters in the lower row. *How many red counters? How many green counters? How many counters altogether?*
▶ Similarly with other combinations involving 5 (5 and 4, 5 and 1, 5 and 2, 5 and 5).
▶ Flash an empty ten frame with 5 red counters in the upper row, and 2 green counters in lower row? *How many red counters? How many green counters? How many counters altogether?*
▶ Similarly with other combinations involving 5 (5 and 2, 5 and 3, 5 and 4, 5 and 5).

Purpose, Teaching and Children's Responses

▶ As above, important goals are for children to ascribe number to flashed patterns and to reason about combinations without counting from one or counting-on.

7.4.4: Using 5 to Partition Numbers in the Range 6 to 10

▶ Place out an empty ten frame. Place 5 red counters in the upper row, and 2 red counters in lower row. *How many counters altogether? How many counters in the upper row? How many counters in the lower row?*
▶ Similarly with other combinations involving 5 (5 and 1, 5 and 3, 5 and 4, 5 and 5).
▶ Flash an empty ten frame with 5 red counters in the upper row, and 4 red counters in lower row. *How many counters altogether? How many counters in the upper row? How many counters in the lower row?*
▶ Similarly with other combinations involving 5 (5 and 1, 5 and 2, 5 and 3, 5 and 5).

Purpose, Teaching and Children's Responses

▶ As above, important goals are for children to ascribe number to flashed patterns and to reason about partitions without counting from one or counting-on.

7.4.5: Combining Numbers to 10

▶ Flash an empty ten frame with 5 red counters in the upper row, 3 red in the lower row, and 2 green in the lower row. *How many red counters? How many green counters? How many counters altogether?*

▶ Similarly with other combinations to 10 (9 and 1, 1 and 9, and so on).

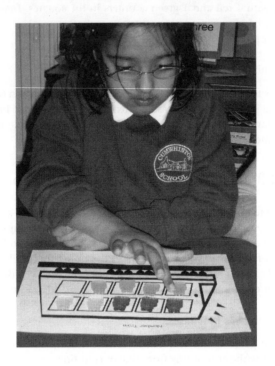

Combining numbers to 10

Purpose, Teaching and Children's Responses

▶ As above, important goals are for children to ascribe number to flashed patterns and to reason about combinations without counting from one or counting-on.

7.4.6: Partitioning 10

▶ Flash an empty ten frame. *How many squares altogether?*

▶ *I'm going to put on 8 red counters? How many empty squares will there be?* Place on 8 red counters. *Watch to see if you were correct.* Flash the ten frame. *Were you correct? How many squares altogether? How many counters? How many empty squares?*

▶ Similarly with other partitions of 10 (9 and 1, 1 and 9, and so on).

Purpose, Teaching and Children's Responses

▶ As above, important goals are for children to ascribe number to flashed patterns and to reason about partitions without counting from one or counting-on.

VOCABULARY

Five frame, ten frame, upper row, lower row.

MATERIALS

Empty five frame.
Five frames showing numbers 1–5.
Empty ten frame.
Ten frame showing numbers 1–10.

ACKNOWLEDGMENT

These activities were adapted from the work of Paul Cobb and colleagues.

Key Topic 7.5: Partitioning and Combining Numbers in the Range 1 to 10

Purpose: To develop facility with combining and partitioning numbers in the range 1 to 10.

LINKS TO LFIN

Main links: Part C, Base-five.
Other links: Part C, Combining and Partitioning Part C, Finger Patterns. SEAL, Stage 3; FNWS and
BNWS, Levels 1–3.

TEACHING PROCEDURES

7.5.1: Describing and Recording Partitions of a Number

▶ Place out a pair-wise ten frame for 6. *How many dots are there on this frame?*
▶ *Tell me two numbers that make 6. Can you see those two numbers on the frame? Show me where the two
numbers are?*
▶ *Can you tell me another two numbers that make 6?*
▶ *Try to make six as many different ways that you can.*
▶ *I am going to write all of the ways of making six.* Use an appropriate notation system to record the
partitions of 6 (see below).
▶ Similarly for other numbers in the range 2 to 10 using both pair-wise and five-wise patterns.

Purpose, Teaching and Children's Responses

▶ Partitions of a number can be recorded as follows –

▶ An important realization for children is that the partitions of 6 (and other numbers) can be organized systematically (for example, 5 and 1, 4 and 2, 3 and 3, 2 and 4, 1 and 5).

▶ Recording could include writing the following arithmetic expressions: 5 + 1; 1 + 5; 4 + 2; 2 + 4; 3 + 3.

▶ 6 + 0 and 0 + 6 could be recorded also, and described as two numbers that also add to 6 but do not involve a partition of 6.

▶ Children could be taught to use the term 'partition' (for example, *make all of the partitions of 5*).

7.5.2: Partitioning Using Flashed Ten Frames

▶ Flash a ten frame showing a pair-wise 6. *How many dots did you see? Tell me two numbers on the frame that make six.*

▶ Flash the ten frame again. *Tell me another two numbers that make six. What two numbers did you see this time?*

▶ Similarly for other numbers in the range 2 to 10 using both pair-wise and five-wise patterns.

Purpose, Teaching and Children's Responses

▶ Encourage children to visualize the partitions and to reason with their visualized partitions.

▶ Children might use counting-by-ones.

▶ Encourage children to use strategies other than counting-by-ones.

▶ Encourage children to reason about different partitions and to make connections among the partitions, for example, 5 and 1 can be changed to 4 and 2, by taking 1 from the 5.

7.5.3: Partitioning and Recording Using Flashed Ten Frames

▶ Flash a ten frame showing a five-wise 8. *How many dots did you see? Tell me two numbers that make 8. Write the two numbers.*

▶ Flash the ten frame again. *Write another two numbers that make 8.*

▶ *Try to write all of the ways we can make 8.*

▶ Similarly for other numbers in the range 2 to 10 using both pair-wise and five-wise patterns.

Purpose, Teaching and Children's Responses

▶ Encourage children to determine associations between different partitions.

▶ For example, 5 and 3 is associated with 6 and 2, by decreasing 6 by 1 and increasing 2 by 1. Similarly 6 and 2 is associated with 4 and 4, 7 and 1 is associated with 1 and 7, and so on.

▶ Partitions can be recorded using the notation or writing arithmetic expressions shown above (see 7.5.1).

7.5.4: Combining Two Numbers Using Visible, Pair-Wise Frames

▶ Place out a ten frame showing a pair-wise 4 and a ten frame showing a pair-wise 3. *How many dots are there on this frame? How many dots on this frame? How many dots altogether? Explain how you know there are seven dots?*

▶ Similarly for other pairs of addends with sum less than or equal to 10 (6 and 3, 1 and 7, 4 and 5 and so on).

Purpose, Teaching and Children's Responses

▶ Discourage children from counting-by-ones to figure out how many on each frame and how many althogther.

▶ Encourage children to give explanations in terms of reorganizing configurations of dots.

7.5.5: Combining Two Numbers Using Visible, Five-Wise Frames

▶ Place out a ten frame showing a five-wise 7 and a ten frame showing a five-wise 2. *How many dots are there on this frame? How many dots on this frame? How many dots altogether? Explain how you know there are nine dots?*

▶ Similarly for other pairs of addends with sum less than or equal to 10.

Purpose, Teaching and Children's Responses

▶ As for 7.5.4.

7.5.6: Combining Two Numbers Using Flashed Frames

▶ Place out a ten frame showing a pair-wise 6. *How many dots are there on this frame?* Then flash a pair-wise 2. *How many dots are there on this frame? How many dots altogether? Explain how you know there are eight dots?*

▶ Similarly for other pairs of addends with sum less than or equal to 10.

▶ Flash a ten frame showing a pair-wise 5. *How many dots are there on this frame?* Then flash a pair-wise 3. *How many dots are there on this frame? How many dots altogether? Explain how you know there are eight dots?*

▶ Similarly for other pairs of addends with sum less than or equal to 10.

Purpose, Teaching and Children's Responses

▶ Encourage children to give explanations in terms of reorganizing configurations of dots.

▶ Encourage reasoning about associations between combinations such as 4 + 1, 4 + 2, 4 + 3, and so on.

▶ Children's reasoning might lead to a generalization such as, increasing the second addend by one increases the sum by one.

▶ As a general rule, use frames of the same kind for each addend.

7.5.7: Combining Two Numbers Using Flashed Frames and Recording

▶ Flash a ten frame showing a pair-wise 5. *How many dots are there on this frame?* Then flash a pair-wise 2. *How many dots are there on this frame? How many dots altogether? Explain how you know there are seven dots? Let's write something to show that five and two make seven.* Use an appropriate recording method to show the combination of 5 and 2.

▶ Similarly for other pairs of addends with sum less than or equal to 10 using two pair-wise frames or two five-wise frames.

Purpose, Teaching and Children's Responses

▶   Combinations can be recorded as follows –

VOCABULARY

Ten frame, pair-wise, five-wise, upper row, lower row.

MATERIALS

Ten frames showing pair-wise patterns for 1 to 10.
Ten frames showing five-wise patterns for 1 to 10.

ACKNOWLEDGMENT

These activities were adapted from the work of Paul Cobb and colleagues.

Key Topic 7.6: Early Multiplication and Division

Purpose: To develop early multiplicative and divisional strategies.

LINKS TO LFIN

Main links: Part D, Early Multiplication and Division.
Other links: FNWS and BNWS, Levels 1–4; SEAL, Stage 1.

TEACHING PROCEDURES

7.6.1: Combining and Counting Equal Groups

▶   Place out ten 2-dot cards. *Show me 2 dots. Now 2 more dots. How many is that altogether? Now 2 more dots. How many is that altogether? Can you put them into a pattern?* (that is, make a 2 × 10 rectangular array of dots) *Now 2 more. How many is that?*
▶   *Watch me as I count the dots. Two, four, six, eight. Count the dots with me by twos. Ready, two, four, six, eight.*
▶   Continue to 10 or 12.
▶   Place out five 3-dot cards. *Show me 3 dots. Show me 3 more. How many is that altogether? Show me 3 more. Can you put them into a pattern?*
▶   *Watch me as I count the dots. Three, six, nine. Count the dots with me by threes. Ready. Three, six, nine.*
▶   Similarly with 5-dot cards.

Combining and counting equal groups (1)

Combining and counting equal groups (2)

Purpose, Teaching and Children's Responses

▶ Children might initially count from one, and count by ones, or count-on by ones from the first group or a subsequent group.

▶ Encourage children to count by 2s, 3s, and so on according to the group size.

Combining and counting equal groups (3)

7.6.2: Determining the Number of Equal Groups

▶ *Here are 10 eggs. Here are some baskets. Each basket should get 2 eggs. Put 2 eggs in each basket. How many baskets have two eggs?*

▶ *Here are 15 biscuits. Here are some plates. Put 3 biscuits on each plate. How many plates get 3 biscuits?*

▶ Similarly with 12 and 4, 20 and 5.

Purpose, Teaching and Children's Responses

▶ Children might count by ones (one, two; one, two; and so on) to make each equal group or might say the number of each equal group without counting-by-ones (two, two, and so on).

▶ Encourage children to work with whole groups when reorganizing the items, and to use the appropriate number name for each whole group rather than referring to each individual item.

7.6.3: Determining the Number in an Equal Share

▶ *Here are 6 counters. Here are 3 people. Can you share the counters out so that each person gets an equal share? How many counters does each person get?*

▶ *Here are 12 counters. Here are 4 people. Can you share the counters out so that each person gets an equal share? How many counters does each person get?*

▶ Similarly with 10 and 2, 12 and 6.

Purpose, Teaching and Children's Responses

▶ Encourage children to use a systematic sharing strategy.

▶ Children might count by ones to confirm the number in each share or confirm the number in each share without counting-by-ones (for example, two, two, two).

Determining the number in equal share

7.6.4: Describing Visible Arrays

▶   Place out a 4 × 6 array. *Here is an array. What do you notice? These are called rows. How many rows are there? What can you say about each row? These are called columns. How many columns are there? What can you say about each column?*
▶   Similarly with the following arrays: 3 × 5, 3 × 2, 6 × 3, and so on.

Describing visible arrays

Purpose, Teaching and Children's Responses

▶ Try to ensure that children have a clear knowledge of the terms 'row' and 'column'.

▶ Encourage children to realize that each row has the same number of items and each column has the same number of items.

▶ Children might say that each row/column has 4/6 dots without counting-by-ones or might count by ones to determine the number in each row/column, or to confirm that each row/column has the same number.

7.6.5: Building Visible Arrays

▶ Place out 3 rows of 6 dots. *Here are some rows. What do you notice about each row? Can you make an array?*

▶ *Point to each of the rows in your array. How many rows are there? Point to each column in your array. How many columns are there?*

▶ Similarly 4 rows of 2 dots, 2 rows of 7 dots, 5 rows of 4 dots, and so on.

Purpose, Teaching and Children's Responses

▶ When the child uses rows to build an array they might have difficulty in recognizing the columns of the array.

▶ Encourage children to see that each row has the same number of dots and each column has the same number of dots.

7.6.6: Determining the Number of Dots on Visible Arrays

▶ Place out a 2 × 5 array. *This array has two rows of 5 dots. How many dots are there altogether?*

▶ Similarly with the following arrays: 2 × 3, 3 × 5, 6 × 2, 4 × 4, and so on.

Purpose, Teaching and Children's Responses

▶ Children might use rhythmic counting – one, *two*, three, *four*, five, *six*, and so on (that is, emphasizing every second number word).

▶ Children might use a combination of counting in multiples and counting-on (2, 4, 6, 8, 10, 11, 12).

▶ Encourage children to count by twos, threes, and so on to determine the number of dots altogether.

VOCABULARY

Two, four, six, eight, ... , three, six, nine, twelve, ... , five, ten, fifteen, twenty, array, row, column, share, sharing, equal groups.

MATERIALS

Dot cards for numbers 2–5.
Counters, small plastic dolls.
Baskets, plates, and so on for equal shares.

Small plastic objects (for example, eggs, cookies) for sharing.

Arrays (for example, 4 × 6, 3 × 5, 3 × 2, 6 × 3, 2 × 3, 6 × 2, 4 × 4). Rows of dots (for example, 3 rows of 6, 4 rows of 2, 2 rows of 7, 5 rows of 4).

EXAMPLES OF WHOLE-CLASS LESSONS DESIGNED FOR THE CHILD AT THE FIGURATIVE STAGE

Lesson 7(a)	**Developing counting-on and counting-back strategies**
Lesson 7(b)	**Row tasks to promote counting-on and counting-back strategies**
Lesson 7(c)	**Using five: developing addition and subtraction pairs to five**
Lesson 7(d)	**Five plus: developing five plus strategies**
Lesson 7(e)	**Numbers to 100**
Lesson 7(f)	**Arrays**
Lesson 7(g)	**Clear the Board**

LESSON 7(A)

Title:	**Developing counting-on and counting-back strategies**
Purpose:	To assist children to develop more sophisticated addition and subtraction strategies.
Links to Key Topics:	7.3
Materials:	For Addo and Subo: spinners or dice, cups, counters, game board.
	For Addition cards: double-sided addition cards (shown below), record sheet.
Introductory Activities:	▶ Teacher asks 5 children to hide behind a screen (large cardboard boxes, for example, television, refrigerator cartons, and so on).
	There are five children hiding behind this screen.
	Teacher asks three children to hide behind another screen.
	There are three children hiding behind this screen. How many children are hiding altogether?
	▶ Repeat this task for subtraction. Have seven children hiding behind the cardboard screen.
	There are seven children hiding behind the screen.
	Select two children to appear. *Two children have come out. How many are left behind the screen?*
Main Focus:	▶ Briefly display and then screen 6 red counters. *Here are 6 red counters.* Briefly display and then screen 4 blue counters. *Here are 4 blue counters. How many counters in all?*
	▶ Briefly display and then screen 9 red counters. *Here are 9 red counters.* Ask the children to look away while screening 3 blue counters. *With the blue counters there are 12 counters altogether. How many blue counters are there?*

▶ Briefly display and then screen 10 red counters. Ask the children to look away while removing 3 counters. Briefly display and then screen the 3 counters. *There were 10 counters and I removed 3. How many counters are left?*

Group/Individual Tasks: Any of these activities would support the main focus of this lesson.

▶ Addo game (Pairs)

First player:

- Spins the red spinner (or throws red die) and places that number of red counters into a cup.

- Spins the green spinner (or throws the green die) and places this number of green counters into the cup.

- Works out how many counters are in the cup.

Second player checks this result by counting the number of counters in cup.

If correct, the first player moves his/her marker this number of spaces around the game board.

▶ Subo game (Pairs)

This is played the same as Addo except only the red counters are used.

The first player:

- Spins the red spinner (or throws the red die) and places this number of counters into the cup.

- The green spinner is then spun (or green die thrown) and this number of counters is removed from the cup.

- This player then works out how many counters remain in the cup.

The second player checks this result by counting the number of counters in cup.

If correct, the first player moves his/her marker this number of spaces around the game board.

Notes for both Addo and Subo:

The numerals on the spinners or dice can be modified to suit the learning needs of the students.

The numerals on the first spinner (or die) are larger than the numerals on the green spinner (or die). Typically the red spinner has 4, 5, 6, 7, 8 and 9 and the green spinner has 1, 2, and 3.

▶ Addition cards (pairs or individual)

Material: Addition cards – on one half there is a numeral and on the other half up to four dots.

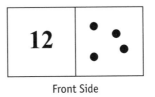

Front Side

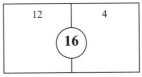

Reverse Side

- Student selects a card and works out the number of dots there are altogether.

- The answer is checked on the reverse side of the card.

- If correct, the student Places this card in the correct stack. If incorrect, places it in the incorrect or 'Needs More Thinking' stack. This is left for the teacher to examine later or children could be directed to write the tasks and their solutions on a record sheet and shown to the teacher after the session.

Covered Dots	Uncovered Dots	Dots Altogether

Teacher gives some verbal tasks to check students' addition and subtraction strategies.

Conclusion/Summary:

▶ *I have five apples and I get two more. How many do I now have?*

▶ *From a pile of ten counters I take three. How many are left?*

LESSON 7(B)

Title:	Row tasks
Purpose:	To promote counting-on and counting-back strategies.
Links to Key Topics:	7.2, 7.3
Materials:	Counters and cover for overhead projector; 'Find the Number' sheet.
Introductory Activities:	None
Main Focus:	▶ Partly screen a row of counters as shown below.

Point to the first visible counter. *This is 14.*

Point to the next counter. *What number is this?*

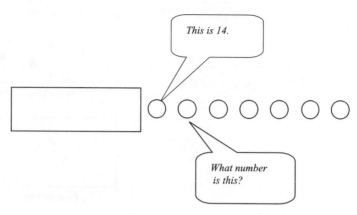

▶ Partly screen a row of counters.

Point to the first visible counter. *This is 36.*

Point to the counter before. *What number is this?*

▶ Repeat tasks asking for numbers two or three away.

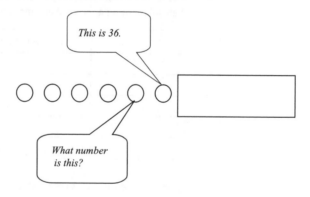

Provide a range of similar tasks.

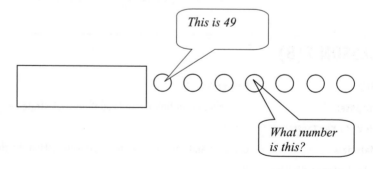

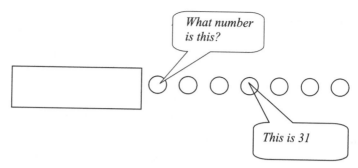

Group/Individual Tasks: ▶ Row task work sheets.

Children write in the missing numbers on this sheet.

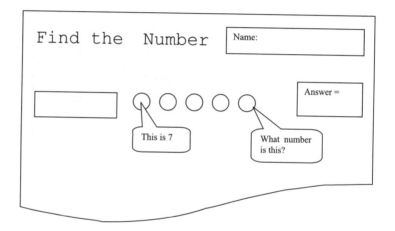

Conclusion/Summary: ▶ Children discuss some of their solutions from the tasks on the 'Find the Number' sheet.

▶ Some more row tasks could be given using the overhead projector.

LESSON 7(C)

Title: **Using five**

Purpose: To develop the addition and subtraction pairs to five.

Links to Key Topics: 7.4

Materials: Five frame for overhead projector; blocks; Fives Snap or Fish cards; fives towers tracking sheet.

Introductory Activities: ▶   Flash a blank five frame. *What did you see?*

         ▶   Flash three red and one two blue counters in a five frame. *What did you see?*
- Flash different arrangements of the red and blue counters.
- Flash four opaque counters (or transparent counters of one colour) in a five frame.
- *What did you see?*
- How many empty spaces were there?

Main Focus: Towers to five

         ▶   Have children make towers five blocks tall using blocks of two colours (for example, red and blue). After experimenting, include the rule that all the red blocks are to be together and all the blue blocks are to be together. The task is to make as many different combinations as possible. These can be listed on a tracking sheet.

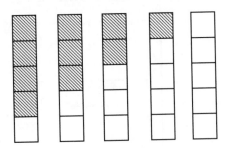

Five Towers

         ▶   The teacher guides the children in a discussion of their results.

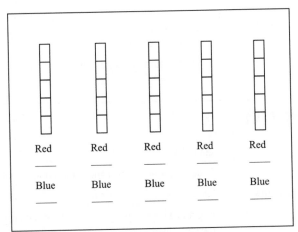

Five Towers Tracking Sheet

Group/Individual Tasks: ▶   Fives Snap.

         Played like traditional Snap where players match fives cards and numeral cards.

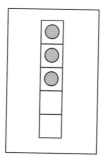

▶ Fives Fish.

Using the cards shown above for Fives Snap, children play Fish. Pairs are made by matching a five card with a numeral card.

Conclusion/Summary: Linking to a different setting.

(a) Screen four counters on the overhead projector (or hide four items in a box). *There are four counters under the screen. How many more do I need to make five counters?*

(b) Screen three counters. There are three counters under the screen. Display another two counters. *Here are another two counters. How many are there altogether? How did you get your answer?*

LESSON 7(D)

Title: **Five plus**

Purpose: Developing five plus strategies.

Links to Key Topics: 7.4, 7.5

Materials: Ten frames for the overhead projector; ten bus for the overhead projector; counters; five plus bingo cards; Five Plus Snap and Fish cards;

Introductory Activities: ▶ Five plus flashes.

Using a tens frame flash five counters (same color) in the top row. *What did you see?*

Flash five plus combinations like those shown below. *Make these on your fingers.*

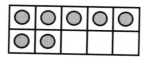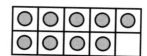

Main Focus: Using the tens bus.

– The full tens bus is displayed then screened (drives behind a hill or building).

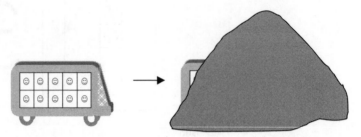

Two people from the top row get off the bus. How many are left? After discussion the bus is displayed with the eight people on board.

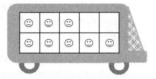

▶ The tens bus with five passengers is displayed then screened.

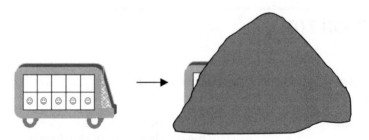

Two counters (people) are placed next to the screen. *If two more get on the bus how many people will now be on the bus?*

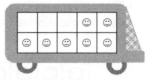

After discussion the bus is displayed with the seven people on board.

Group/Individual Tasks: ▶ Give other tasks with people getting on and off the bus involving five and ten.

All these activities are designed to support the development of ten plus combinations.

▶ Five Plus Bingo.

Children have playing boards as shown below and some counters. The teacher or leader rolls a cube with numerals 0, 1, 2, 3, 4, 5 and calls out the number that is face-up. A traditional die can be used with the teacher calling out 0 when the 6 is thrown. Students add this number to five and place a counter on one of the squares containing this number. The aim is to get four (or five) counters in a row.

9	8	1	5	8	6	9	
6	7	1		9	7	8	5
7	9	8	6	1		5	6
5	8		7	6	9	9	6
9	1	5	9	7		6	8
1	6		5	8	6	9	7
8		9	6	9	7	5	1
7	9	8	5	1	6		8

▶ Five Plus Snap.

Snap is played with five Plus cards and numeral cards.

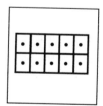

▶ Five Plus Fish.

Using the cards shown above for Five Plus Snap, students play Fish. Pairs are made by matching a five plus card with a numeral card.

▶ How Many More To Make Ten?

Pairs of students have a stack of cards with five plus patterns on one side and the matching numeral on the reverse side. The stack of cards is placed with the numerals face-up. Players in turn select the top card and work out how many more to make ten. The result can be checked by looking at the reverse side of the card. If correct (for example if the card displayed 6 and the student's answer was four), the student places four counters on the score sheet.

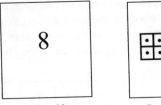

Facing side   Reverse side

Conclusion/Summary: Using a different setting for five plus combinations.

▶ Screen five counters on the overhead projector (or hide five items in a box). *There are five counters under the screen. How many more do I need to make seven counters?*

    – Screen five counters. There are five counters under the screen. Display another two counters. *Here are another two counters. How many are there altogether? How did you get your answer?*

LESSON 7 (E)

Title: **Numbers to 100**

Purpose: To provide children with activities that assist them in learning the numerals to 100 and forward and backward number sequences to 100.

Links to Key Topics: 7.1, 7.2

Materials: Large numeral cards which are chosen to suit the needs of the class or group; small numeral cards for Triads game and for numeral arranging task.

Introductory Activities: Smallest to Biggest.

▶ Each member of the class is given a numeral card. Their task is to arrange themselves into ascending order.

▶ Alternatively, groups of children could be given the same range of cards with the aim of being the first to arrange themselves into ascending or descending order as determined by the teacher.

Main Focus: Jimbo.

A sequence of numeral cards is used for this task (for example, 48 to 61).

▶ The teacher holds the stack of cards and invites a child to choose one and place it on the chalkboard ledge or peg it on a line.

▶ Another child is chosen to select a card and place it in relation to the existing card. This procedure continues until all the cards are used.

▶ As each card is placed on the ledge, the class has the opportunity to discuss its placement.

- ▶ If children have difficulty placing a card, they can ask a friend to come and assist them.

- ▶ When all of the cards are placed, the class can say the sequence forward and backwards.

- ▶ The teacher asks the class to close their eyes and a card is turned over. *What card is turned over? How do you know?*

- ▶ Additional cards are turned over and these questions repeated.

- ▶ All of the cards are turned over so no numerals are seen. The teacher shows one card and then points to other cards each time asking: *What number is this?*

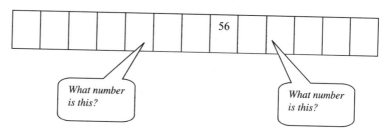

Large numeral cards – multiples of ten

The previous task can be repeated using multiples of ten.

Group/Individual Tasks: Triads.

Cards are selected containing numerals in an appropriate range for the children. For example the cards 10 to 60 might be chosen. If the teacher wants children to focus on a smaller sequence of numerals (for example, the teens) then up to four cards containing each teen numeral could be used.

This game is played like traditional Fish. Instead of making pairs, children make runs of three cards (three consecutive numbers). Variations include:

- ▶ allowing players to add to their own existing cards to make larger runs;

- ▶ allowing players to add to existing runs of other players with the aim of being the first to have no cards left in their hands;

- ▶ cards containing multiples of ten can be used.

Smallest to Biggest.

Individuals, pairs or small groups of children have a group of numeral cards that they arrange on the floor from smallest to largest.

Conclusion/Summary: Either of these activities would make a suitable conclusion to this lesson.

- ▶ Number after Bingo.

   In this variation of the Bingo children place a counter on their Bingo board if they have the number after the number called out by the teacher.

- ▶ Number before Bingo.

   Children need to get the number before the one called out by the teacher.

LESSON 7 (F)

Title:	**Arrays**
Purpose:	Arrays are used to promote the counting of equal groups and to develop early multiplication strategies.
Links to Key Topics:	7.6
Materials:	Hundred square for the overhead projector; counters; arrays for the overhead projector; array Fish cards

Introductory Activities:

▶ Using a hundred square on the overhead projector, the teacher places a transparent counter on the multiples of five.

▶ As the teacher points to each counter, the class says the number.

▶ The transparent counters are replaced with opaque counters and the procedure repeated.

Main Focus:

▶ Counting equal groups.

Three groups of five counters are arranged on the overhead projector.

How many counters are there? How did you find your answer?

I counted one, two, three ... to fifteen.

Can we find a quick way of finding how many counters there are?

Count by fives; five, ten, fifteen – this is much quicker.

▶ Linking to the complementary division task.

If there are three children and we divide the fifteen counters between them, how many will each child get? How can we find the answer?

Just keep giving each child a counter until all fifteen are used up. We know that if we count by fives we get fifteen ... five, ten, fifteen ... so I think each person gets five.

▶ Using arrays.

Display a 3 × 5 array.

What can you see?

Lots of dots in rows.

I can see fifteen ... I counted them.

There are three rows and each row has five.

I saw five up and down rows and there are three in each row.

How can we find out how many dots there are altogether? Try to find any fast ways of working out how many there are.

We can use our fives to count how many dots. Five, ten, fifteen. There must be fifteen.

▶ Other array tasks are given.

Group/Individual Tasks: Array Fish.

The teacher should choose arrays based on the multiples used in the main focus section of this lesson.

▶ This is played the same as traditional Fish. Screened array cards are matched with numeral cards.
  – Array and numeral cards such as those shown below for multiples of five are used.
  – The teacher may choose to have children focus on particular multiples (for example, fives) or choose cards from a range of multiples (for example, 3, 4 and 5, and so on).

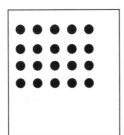

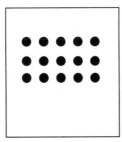

Conclusion/Summary: Fives Race.

▶ Divide the class into three teams (or let the children divide themselves into three teams).

Give each team member a large multiple of five.

The teams arrange themselves in order.

Some arrays could be displayed and children asked to find the number of dots.

LESSON 7 (G)

Title:	**Clear the Board**
Purpose:	To assist children to develop addition combinations to 12.
Links to Key Topics:	7.5
Materials:	Clear the Board sheets, counters and dice.
Introductory Activities:	None.

Main Focus: Clear the Board game.

▶ This game assists children to partition numbers.

▶ Each child has a Clear the Board sheet and ten counters (the number of counters can be varied).

▶ Players place all of their counters anywhere on the board (no more than four can be placed in any one column).

▶ The teacher rolls the two dice and tells the class the numbers that show. The children add these numbers together to give a total (for example if a 3 and a 5 show, the total is 8).

▶ One counter is removed from this column (say the 8 column) if a player has counters in this column.

▶ The game proceeds until a player has removed all the counters from his/her board.

▶ The teacher questions children about how they placed their counters.

▶ The game is played again and the teacher again questions the class about their placement of the counters.

Group/Individual Tasks: ▶ The class is divided into pairs and the each pair is given two dice.

▶ The Clear the Board game is now played in pairs with children taking turns to roll the dice, add the numbers and remove counters as described above.

Conclusion/Summary: ▶ With the class assembled, the teacher:
 – questions children about their strategy for placing the counters;
 – rolls two dice with the children adding the two number together;
 – asks the class how many different ways there are of making various numbers.

Clear the Board Playing Sheet

2	3	4	5	6	7	8	9	10	11	12

8
Teaching the Counting-On Child

This chapter focusses on teaching children at the Counting-On and Counting-Down-To Stages on the SEAL (Stages of Early Arithmetical Learning – see Chapter 1). First, a detailed description of the knowledge and strategies typical of these two stages is provided. This is followed by descriptions of six key topics which can provide a basis for teaching the child at the Counting-On or Counting-Down-To Stages. The six key topics consist of a total of 41 teaching procedures.

THE TYPICAL COUNTING-ON CHILD

This section provides an overview of a typical child at the Counting-On Stage (Stage 3), or the Counting-Down-To Stage (Stage 4) of the Stages of Early Arithmetical Learning. Table 8.1 sets out the stage and indicative levels for the Counting-On child on the models pertaining to FNWSs, BNWSs, Numeral Identification and Tens and Ones. The overview discusses the six aspects of early number knowledge listed in Table 8.1.

Table 8.1 Stage and levels of a typical counting-on child

Model	Stage/level
Stage of Early Arithmetical Learning (SEAL)	3/4
Level of Forward Number Word Sequences (FNWSs)	5
Level of Backward Number Word Sequences (BNWSs)	4
Level of Numeral Identification	3
Level of Tens and Ones Knowledge	1
Level of Early Multiplication and Division Knowledge	1

Stage of Early Arithmetical Learning

The child at the Counting-On Stage (Stage 3) has developed one or more of the advanced counting-by-ones strategies of counting-up-from, counting-up-to, and counting-down-from. The child at Stage 4 has developed counting-down-to as well as the other advanced counting-by-ones strategies. Thus children at these stages will solve additive tasks by counting-up-from, and typically count-up-from the first mentioned or larger addend. On additive tasks involving two covered collections children are able to keep track of six or more counts when counting the second collection. In many cases children are able to use these counting-on strategies in the range 1 to 100. Eighty-seven and five for example, is solved by counting-on five from 87. The child knows to stop at 'ninety-two' because they realize they have made five counts. Thus prior to commencing the count from 87, the child anticipates that they can keep track of the number of counts. This anticipation and capacity to keep track of the number of counts is the hallmark of this stage.

Counting-Up-To and Counting-Down-From

The counting-up-from strategy described in the previous paragraph is also referred to simply as counting-on. Counting-on also includes counting-up-to, which typically arises when solving tasks involving a small unknown addend, for example, $8 + x = 11$, presented using counters of two colors (referred to as missing addend tasks). Some children who count-on to solve additive tasks might not have developed the counting-up-to strategy. They might be unfamiliar with missing addend tasks, and it is common for children to misinterpret a missing addend task as an additive task (for example, in the task just described they might attempt to start from eight and count-on eleven). In the case of most children though, counting-up-to emerges at the same time or soon after counting-up-from. At around the same time children learn to use counting-down-from to solve subtractive tasks with small known differences (referred to as removed items tasks). Thus on a task such as removing three counters from a collection of 14 (with the counters concealed), the child will count down three from 14. As before, the child anticipates that they can keep track of the number of counts, and thus stops after three counts.

Counting-Down-To

Tasks involving small, unknown subtrahends, for example remove some counters from a collection of 11 to leave 8 (known as missing subtrahend tasks), are also appropriate for children at Stage 3 or 4. Tasks of this kind can evoke the counting-down-to strategy. Thus the child counts back from 11 until they reach 8, and keeps track of the number of counts after saying 'eleven', that is, three counts. In the terms of the Learning Framework in Number, counting-down-to is regarded as more sophisticated than the other advanced counting-by-ones strategies. Thus it is not surprising to find children who can solve additive, missing addend, and removed items tasks like those described above, but cannot solve missing subtrahend tasks of the kind described, that is, they cannot use a counting-down-to strategy. At the same time it is not considered crucial to focus teaching on the development of counting-down-to. Rather, when children have developed robust counting-on and counting-back strategies, teaching should focus on advancing the child to the facile stage (that is, Stage 5).

FNWSs, BNWSs and Numeral Identification

It is typical for the child at this stage to be facile with FNWSs to 100 and beyond. In the case of BNWSs children might be facile to beyond 30 but not necessarily to 100. A reasonably common mistake in the case of saying BNWSs is the following: when saying the words backward from 53 for example, the child says '53, 52, 51, 40, 49, 48 … '. This kind of mistake occurs at any decade number and a likely explanation is that the child counts backwards by ten from the decade number in order to determine the next decade (for example, counts '50, 40' to figure out the decade number before the fifties). Children at this stage typically can recognize and identify most or all of the numerals in the range 1 to 100 and beyond, although some digit reversal errors (for example, '27' is named as 'seventy-two') may persist. Children might also be able to name some 3-digit numerals. The 3-digit numerals with a zero in the ones or tens (for example, 620, 407) are less likely to be correctly identified.

Tens and Ones

The child at this stage typically has made little progress in developing knowledge of tens and ones. They are likely to be able to count forwards and backwards by tens on the decade (10, 20, 30, and so on) but not off the decade (2, 12, 22, and so on). On tasks involving tens and ones materials (for example, bundling sticks) they might increment by tens in the case of whole tens and no ones (for example, 10, 20, 30, and so on), but often this amounts to no more than recognition of the familiar sequence of the decade number words (ten, twenty, and so on). In cases involving incrementing by tens off the decade or incrementing by tens and ones, these children typically resort to counting-on by ones, or they incorrectly count ones as tens and so on.

Early Multiplication and Division

The child at this stage might also have developed initial multiplication and division knowledge relating to equal groups, equal shares and arrays, for example, combining equal groups and finding the total, sharing equally and finding the number in one share, making an array or determining how many dots in an array.

The Way Forward

The child at this stage has relatively facile knowledge of FNWSs and BNWSs by ones and is ready to learn sequences by 2s, 10s, 5s, 3s and 4s. The child might well already have some knowledge of counting forwards by 2s, 10s and 5s, and this knowledge can be consolidated and extended, and counting backwards by 2s and so on can also be developed. As well, the child at this stage is ready to learn to recognize and identify 3-digit numerals. The extension of the number naming system to 3-digit numbers follows a consistent and regular pattern, and children find this relatively easy to learn, although some children might encounter some initial difficulty in learning to identify the 3-digit numerals with a zero in the ones or tens (for example, 620, 407). The child as this stage is also ready to begin learning about tens and ones. This learning can include incrementing and decrementing by tens on and off the decade and by tens and ones. The child should also learn to add and subtract numbers in the range 1 to 9, to and from a decade number (for example, 40 + 6, 37 − 7, 48 + x = 50, 60 − x = 57), and to add and subtract in the range 1 to 20 using grouping by 5 and grouping by 10. Finally, the child can learn to use strategies involving skip counting, repeated addition and so on, to solve multiplicative and divisional tasks involving equal groups and arrays.

The next section contains six key topics and 41 teaching procedures which form the basis of an appropriate instructional programme for the counting-on child.

Key Topic 8.1: Number Word Sequences by 2s, 10s, 5s, 3s and 4s

Purpose: To develop facility with forward and backward number word sequences by 2s, 10s, 5s, 3s, and 4s in the range 1 to 100.

LINKS TO LFIN

Main links: Part D, Multiplication and Division.
Other links: Part A, Tens and Ones; FNWS and BNWS, Levels 1–5.

TEACHING PROCEDURES

8.1.1: FNWSs by 2s, from 2

▶ *Count with me from 1. We'll take turns to say a number. I'll say 1, you say 2, I'll say 3, and you say 4 and so on. Ready, 1, 3, … 19.*
▶ *This time, I'll say my number very softly, and you say your number loudly. Ready, 1, 3, … 19.*
▶ *This time when it's my turn I will nod but I won't say my number. Ready.* Nod to indicate start and continue nodding after each count.
▶ *That's called counting by twos. Count by twos again. Ready, go!*

Purpose, Teaching and Children's Responses

▶ The activities in this teaching procedure focus on number word sequences of multiples. As for Key Topic 7.1 these activities involve the number words alone. These activities can be complemented by activities involving number word sequences of multiples and repeated equal groups of items (see other teaching procedures in this key topic).
▶ Children can use number word sequences of multiples to solve tasks involving repeated equal groups.
▶ These activities can result in children saying the alternate number words (that is, 1, 3, 5, and so on) subvocally and reflecting on number word sequences.

8.1.2: FNWSs and BNWSs by 2s

▶ *This time I am going to use the 2-dot cards. Count by twos as I move the cards. Ready, go!* Move the 2-dot cards slowly, one at a time (from 2 to 10).
▶ *This time, I'm going to take the cards away. Count backwards by twos as I take the cards away. Ready, go!* Remove the 2-dot cards slowly one at a time (from 10 to 2).
▶ *Now count backwards by twos from ten, without the cards. Ready, go!*
▶ *This time, count forwards by twos to 20. Ready, go! Now count backwards by twos from 20. Ready, go!*

Purpose, Teaching and Children's Responses

▶ The 2-dot cards constitute a setting where the child's attention can focus on items arranged in equal groups.
▶ In these situations children can come to regard each group as a unit (that is, a unit of two) rather than as a composite (that is, consisting of 2 ones).
▶ Counting backwards by twos can help to strengthen knowledge of the number word sequence of multiples of twos (two, four, six, and so on).
▶ Children can use knowledge of backward sequences of multiples to solve a range of number tasks (for example, divisional and subtractive tasks).

8.1.3: FNWSs by 10s, from 10

▶ *I am going to put out the bundles of ten one at a time. Count by tens as I put out the bundles (10, 20, … 100).*
▶ *This time, count by tens without the bundles. Ready, go!*

Purpose, Teaching and Children's Responses

▶ In these situations children can come to regard each group of 10 as a unit (that is, a unit of 10) rather than as a composite (that is, consisting of 10 ones).

▶ Children can use knowledge of FNWSs by 10s to solve additive, subtractive and multiplicative tasks involving 2-digit numbers.

8.1.4: FNWSs and BNWSs by 10s

▶ *I am going to put out the bundles of ten one at a time. Count by tens as I put out the bundles (10, 20, ... 100).*

▶ *This time, count backwards by tens, from 100, as I remove the bundles. Ready, go! (100, 90, ... 10).*

▶ *This time, count forwards and backwards by tens to 50 without the bundles. Ready, go! (10, 20, ... 50, 50, 40, ... 10).*

▶ *This time, count forwards and backwards by tens to 100 without the bundles. Ready, go! (10, 20, ... 100, 100, 90 ... 10).*

Purpose, Teaching and Children's Responses

▶ Children can use knowledge of BNWSs by 10s to solve subtractive, multiplicative and divisional tasks involving 2-digit numbers.

▶ In similar vein to 8.1.1 above, activities with number word sequences alone are complementary to activities involving number word sequences and repeated equal groups of items.

8.1.5: FNWSs by 5s, from 5

▶ *I am going to put out the 5-dot cards. Each time I put out a 5-dot card, you tell me how many dots altogether.* Put out one 5-dot card. *Okay, how many dots?* Put out a second 5-dot card. *How many dots now?* Continue to six 5-dot cards.

▶ *That's called counting to 30 by fives. Ready, count by fives while I move the 5-dot cards.*

▶ *This time, count to 30 by fives without the cards. Ready, go!*

▶ *This time, count to 50 by fives. Ready, go!*

▶ *This time, try to count to 100 by fives.*

Purpose, Teaching and Children's Responses

▶ In similar vein to 8.1.1 and 8.1.2 above, children can use their knowledge of these sequences to solve various tasks and can come to regard each group of 5 as a unit.

▶ 5-dot patterns can be domino, linear (like a five frame), or in a pairs pattern.

8.1.6: FNW5s and BNWSs by 5s

▶ *I am going to use the 5-dot cards. Count by fives as I move the cards. Ready, go!* Move the 5-dot cards slowly, one at a time (from 5 to 30).

▶ *This time, I'm going to take the cards away. Count backwards by fives as I take the cards away. Ready, go!* Remove the 5-dot cards slowly one at a time (from 30 to 5).

▶ *Now count backwards by fives from 30, without the cards. Ready, go!*

▶   *This time, count forwards by fives to 50. Ready, go! Now count backwards by fives from 50. Ready, go!*
▶   *This time, count forwards by fives to 100. Ready, go! Now count backwards by fives from 100. Ready, go!*

Purpose, Teaching and Children's Responses

▶   Some children count backwards by 5s incorrectly as follows, 'ninety, eighty, eighty-five, seventy, seventy-five, and so on'. This is analogous to counting backwards by ones incorrectly as 'forty-two, forty-one, thirty, thirty-nine, thirty-eight, and so on' (see Key Topic 7.1).

8.1.7: FNWSs by 3s, from 3, and by 4s, from 4

▶   *I am going to put out 3-dot cards. Each time I put out a 3-dot card, you tell me how many dots altogether.* Put out one 3-dot card. *Okay, how many dots?* Put out a second 3-dot card. *How many dots now?* Continue to four 3-dot cards.
▶   *That's called counting to 12 by threes? Ready count by threes while I move the 3- dot cards.*
▶   *This time, count to 12 by threes without the cards. Ready, go!*
▶   *This time, count to 18 by threes. Ready, go!*
▶   *This time, try to count to 30 by threes.*
▶   Similarly to 40 by 4s with and without using 4-dot cards.

Purpose, Teaching and Children's Responses

▶   8.1.6 applies similarly here.

8.1.8: FNWSs and BNWSs by 3s and by 4s

▶   *I am going to use the 3-dot cards. Count by threes as I move the cards. Ready, go!* Move the 3-dot cards slowly, one at a time (from 3 to 12).
▶   *This time, I'm going to take the cards away. Count backwards by threes as I take the cards away. Ready, go!* Remove the 3-dot cards slowly one at a time (from 12 to 3).
▶   *Now count backwards by threes from 12, without the cards. Ready, go!*
▶   *This time, count forwards by threes to 18. Ready, go! Now count backwards by threes from 18. Ready, go!*
▶   *This time, count forwards by threes to 30. Ready, go! Now count backwards by threes from 30. Ready, go!*
▶   Similarly forwards and backwards to 40s by 4s.

Purpose, Teaching and Children's Responses

▶   8.1.6 applies similarly here.

VOCABULARY

2-dot cards, 3-dot cards, 4-dot cards, 5-dot cards, bundles of ten, count by twos/threes/fives, count backwards by twos/threes/fives.

MATERIALS

2-dot cards, 5-dot cards, 3-dot cards, 4-dot cards.
Bundling sticks.

ACKNOWLEDGMENT

These activities were developed in the Mathematics Recovery project.

Key Topic 8.2: Numerals from 1 to 1,000

Purpose: To develop knowledge of numerals in the range 1 to 1000.

LINKS TO LFIN

Main links: Numeral Identification, Levels 1–4.
Other links: FNWS and BNWS, Levels 1–5; SEAL, Stages 1–3.

TEACHING PROCEDURES

8.2.1: Sequencing and Naming 100s Cards

▶ Place out the hundreds numeral cards in order. *These numbers are the hundreds. I am going to say them starting from 100.* Point to each numeral card in turn. *One hundred, two hundred, … one thousand. Now you say the numbers, starting from one hundred. Point to each number as you say its name.*
▶ Place out the numeral cards from 100 to 500 in random order. *Put these cards in order starting from 100. Now say these numbers starting from 100, and point to each one as you say it.*
▶ *Similarly with cards from 600 to 1,000 and 100 to 1,000.*
▶ Place out a hundreds numeral track with lids closed. Turn over the lid covering 600. *What number is that? Going by hundreds, what is the next number? Turn over the lid and check to see if you were correct. Were you correct?* Turn over the lid covering 300. *What number is that?* Place a marker beside the lid covering 600. *Going by hundreds, what number is under here? Check it. Were you correct?* And so on.

Purpose, Teaching and Children's Responses

▶ The activities in this teaching procedure help to develop knowledge of number word sequences and numerals.
▶ Children can use knowledge of number word sequences by 100s (one hundred, two hundred, three hundred, and so on) to solve additive and subtractive tasks involving 3-digit numbers.

8.2.2: Naming 3-Digit Numerals Using Arrow Cards

▶   Place out the arrow cards for 200, 40 and 6. *Put the arrow cards together. What number is that?* Similarly with other sets of three arrow cards.

▶   Place out the arrow cards for 700 and 60. *Put the arrow cards together. What number is that?* Similarly with other sets of two arrow cards (for a hundreds number and a tens number).

▶   Place out the arrow cards for 500 and 2. *Put the arrow cards together. What number is that?* Similarly with other sets of two arrow cards (for a hundreds number and a ones number).

Naming 3-digit numerals using arrow cards (1)

Naming 3-digit numerals using arrow cards (2)

Naming 3-digit numerals using arrow cards (3)

Purpose, Teaching and Children's Responses

▶ In similar vein to Key Topic 7.2, children can and should learn the names of 3-digit numerals at a time when they might have little understanding of the place value associated with the numerals. The nomenclature for 3-digit numbers is regular and hence learning the names of 3-digit numbers is relatively easy.

Learning to name 3-digit numerals can serve to reduce the likelihood that children will incorrectly read 2-digit numerals from right to left, for example, reading '72' as 'twenty-seven' because experience in naming 3-digit numerals will support the practice of reading the left-hand digit first, in the case of 2-digit numerals as well as 3-digit numerals.

▶ The 3-digit numerals with '0' in the tens place (for example, 406) or in the ones place (for example, 460) are often more difficult for children to name.

8.2.3: Naming 3-Digit Numerals Using Digit Cards

▶ Place out the digit cards for 3, 7 and 5. *Put the digit cards together to make a number in the hundreds. What number did you make? Rearrange the cards to make another number. What number did you make this time? Now make another number.* And so on.

▶ Place out the digit cards for 6, 2 and 0. *Put the digit cards together to make a number in the hundreds. What number did you make? Rearrange the cards to make another number in the hundreds. What number did you make this time?*

Purpose, Teaching and Children's Responses

▶ It is important to give children sufficient time to think and to reflect on their thinking.
▶ These activities can help to develop specific strategies for naming 3-digit numbers.

8.2.4: Sequencing and Naming Decade Numerals beyond 100

▶ Place out numeral cards for decade numbers from 610 to 700, in random order. *Put these cards in order from left to right. Read the numbers from left to right.* Similarly for other hundreds.
▶ Place out numeral cards for decade numbers from 360 to 450, in random order. *Put these cards in order from left to right. Read the numbers from left to right.* Similarly for other sequences of 10 decade cards.

Purpose, Teaching and Children's Responses

▶ These activities can form a basis for incrementing 3-digit numbers by 10.

8.2.5: Sequencing 3-Digit Numerals

▶ Place out the following numeral cards; 207, 217, ... 297, in random order. *Put these cards in order from left to right. Read the numbers from left to right.*
▶ Similarly for sets of numeral cards such as: 362, 372, ... 452; 46, 146, ... 946; 82, 92, ... 212; 11, 111, 211, ... 911. And so on.

Purpose, Teaching and Children's Responses

▶ When children read these numerals sequences they can become aware of patterns in the corresponding number word sequences.
▶ In the case of sequences of 3-digit numerals increasing by 10, children might have difficulty reading sequences starting from a ones or teens number (for example, 207, 217, 227, and so on; 412, 422, 432, and so on ...) because of the irregularities in the name of the teen part of the number).

8.2.6: Ordering 3-Digit Numerals

▶ Place out the following four numeral cards, 98, 201, 410, 905, in random order. *Put these cards in order from left to right. Read the numbers from left to right.*
▶ Similarly for other sets of numeral cards.

Purpose, Teaching and Children's Responses

▶ Most of the notes in Key Topic 7.2: Ordering 2-Digit Numerals, also apply to this teaching procedure.

8.2.7: Sequences of Multiples with Numerals

▶ Place out a numeral track with the following sequence of numerals: 2, 4, 6, ... 20. Turn the lids over to uncover the numerals on the numeral track.
▶ *Let's say these numbers together – two, four, six, ... twenty.*
▶ *Now let's say them backwards – twenty eighteen, sixteen, ... two.*
▶ Cover the numerals with the lids. *Now say the numbers starting from two, and turn over each lid after you say the number.*
▶ Cover the numerals with the lids. *Now say the numbers starting from twenty and going backwards. Turn over each lid after you say the number.*
▶ Similarly with numeral tracks with the sequence by 3s, 5s and 10s.

Purpose, Teaching and Children's Responses

▶ The activities in this teaching procedure complement those in Key Topic 8.1 above because the latter involve the number words alone.

▶ These activities can strengthen children's knowledge of number sequences. This knowledge can be used to solve various kinds of number tasks including involving repeated equal groups.

VOCABULARY

One hundred, two hundred, … one thousand, arrow cards, digit cards, rearrange, left to right. See also Key Topic 7.2.

MATERIALS

Hundreds numeral cards (100 to 1000).
Hundreds numeral track.
Arrow cards for 1 to 9, 10 to 90 and 100 to 900.
3 × cards for digits from 0 to 9 (three cards for each digit, or magnetic digits, or digit tiles.
Decade numbers in hundreds, for example, 610, 620 to 700 and so on.

ACKNOWLEDGMENT

These activities were developed in the Mathematics Recovery project.

Key Topic 8.3: Incrementing by Tens and Ones

Purpose: To develop the facility to increment and decrement numbers by tens and ones, in the range 1 to 100.

LINKS TO LFIN

Main links: Tens and Ones, Level 2.
Other links: SEAL, Stage 3; FNWS and BNWS, Levels 1–5.

TEACHING PROCEDURES

8.3.1: Incrementing and Decrementing on the Decade, by Tens

▶ Place out 3 bundles (of ten). *How many sticks are there?* Place out another bundle. *How many sticks are there now?* Continue to 10 bundles.

▶ Remove one bundle. *How many sticks are there now?* Remove another ten. *How many sticks are there now?* Continue to zero bundles.

▶ Briefly display and then screen 3 bundles. *How many sticks are there under the screen?* Place another bundle under the screen. *How many now?*

▶   Continue to six bundles. *How many sticks are there under the screen? Check to see if you are correct.*

▶   Place another bundle under the screen. *How many now?* Continue to 10 bundles. *How many sticks now? How many bundles? Check to see if you are correct.*

▶   Remove one bundle. *How many sticks are there now?* Remove another bundle. *How many sticks are there now?* Continue to six bundles. *How many now? Check to see if you are correct.*

▶   Remove one bundle. *How many sticks are there now? How many bundles are there now?* Continue to zero bundles.

▶   Continue with similar examples.

Incrementing and decrementing on the decade

Purpose, Teaching and Children's Responses

▶   The activities in this key topic have the purpose of developing children's beginning knowledge of tens and ones. This is a forerunner to the development of place value knowledge.

▶   An important goal of this teaching procedure is for children to be able to regard a bundle simultaneously as one ten and ten ones, that is both as a unit and as a composite.

8.3.2: Incrementing and Decrementing off the Decade, by Tens

▶   Place out 3 bundles and 2 sticks. *How many sticks are there?* Place out another bundle. *How many sticks are there now?* Continue to 10 bundles and 2 sticks.

▶   Remove one bundle. *How many sticks are there now?* Remove another ten. *How many sticks are there now?* Continue to zero bundles and 2 sticks.

▶   Briefly display and then screen 2 bundles and 4 ones. *How many sticks are there under the screen?* Place another bundle under the screen. *How many now?*

▶ Continue to 6 bundles and 4 ones. *How many sticks are there under the screen? Check to see if you are correct.*

▶ Place another bundle under the screen. *How many now?* Continue to 10 bundles and 4 ones. *How many sticks now? Check to see if you are correct.*

▶ Remove one bundle. *How many sticks are there now?* Remove another bundle. *How many sticks are there now?* Continue to six bundles. *How many now? Check to see if you are correct.*

▶ Remove one bundle. *How many sticks are there now?* Continue to zero bundles and 4 ones.

▶ Continue with similar examples.

Incrementing and decrementing off the decade

Purpose, Teaching and Children's Responses

▶ An important goal of this teaching procedure is for children to be able to increment and decrement by tens in settings involving displayed or concealed collections of tens and ones.

▶ Incrementing off the decade is usually more difficult than incrementing on the decade.

8.3.3: Incrementing by Tens and Ones

▶ Briefly display and then screen 4 bundles. *How many sticks are there under the screen?* Place 2 bundles and 1 stick beside the screen. *How many sticks now?*

▶ Put the 2 bundles and 1 stick under the screen and place out 1 bundle and 3 sticks. *How many sticks now? Check to see if you are correct.*

Incrementing by tens and ones

▶ Briefly display and then screen 1 bundle and 4 sticks. *How many sticks are there under the screen?*
▶ Place 2 bundles beside the screen. *How many sticks now?* Put the 2 bundles under the screen and place out 1 bundle and 3 sticks. *How many sticks now? Check to see if you are correct.*
▶ Place the 1 bundle and 3 sticks under the screen and place out 2 bundles and 1 stick. *How many sticks are there now? Check to see if you are correct.*
▶ Continue with similar examples.

Purpose, Teaching and Children's Responses

▶ An important goal of this teaching procedure is for children to be able to increment flexibly by tens or ones in settings involving displayed or concealed collections of tens and ones.
▶ Typical errors are to increment by ten instead of one and vice versa, for example, given 2 bundles and 3 sticks, the child might say 10, 20, 30, 40, 50.

8.3.4: Decrementing by Tens and Ones

▶ Display 9 bundles and 6 sticks. *How many sticks are there?* Place a screen over the 9 bundles and 6 sticks, and remove 2 bundles. *How many sticks now?*
▶ Remove 1 bundle and 1 stick. *How many sticks now? Check to see if you are correct.*
▶ Remove 2 bundles. *How many sticks now? Check to see if you are correct.*
▶ Remove 1 bundle and 3 sticks. *How many sticks now? Check to see if you are correct.*
▶ Continue with similar examples.

Purpose, Teaching and Children's Responses

▶   An important goal of this teaching procedure is for children to be able to decrement flexibly by tens or ones in settings involving displayed or concealed collections of tens and ones.

VOCABULARY

Bundles of ten, singles.

MATERIALS

Bundling sticks.
Rectangles of cardboard for screening bundles and sticks.

ACKNOWLEDGMENT

These activities were developed in the Mathematics Recovery project.

Key Topic 8.4: Adding and Subtracting to and from Decade Numbers

Purpose: To develop the facility to add numbers in the range 1 to 9, to and from decade numbers, and to subtract numbers in the range 1 to 9, to and from decade numbers.

LINKS TO LFIN

Main links: Tens and Ones, Levels 2–3.
Other links: SEAL, Stage 5; FNWS and BNWS, Levels 1–5.

TEACHING PROCEDURES

8.4.1: Adding from a Decade

▶   Place out 4 ten (10 dot) frames. *How many tens are there? How many dots are there?* Place out a ten (4 dot) frame. *How many dots are there now?*
▶   Briefly display and then screen 3 ten (10 dot) frames. *How many tens are there? How many dots are there?* Place a ten (6 dot) frame under the screen. *How many dots are there now?* Check to see if you are correct.
▶   Continue with similar examples.

Purpose, Teaching and Children's Responses

▶   Developing the facility to add and subtract to and from decade numbers constitutes an important basis for addition and subtraction involving 2-digit numbers.
▶   The activities in this teaching procedure are labeled 'adding from a decade' because the first addend is a decade number.

▶ It is important to provide sufficient time for children to think about the tasks and to reflect on the results of their thinking.

▶ Children might become aware of the semantic link between the names of the two addends and the name of the sum, for example, 'twenty and four make twenty-four'.

8.4.2: Subtracting to a Decade

▶ Place out 7 ten (10 dot) frames and a ten (6 dot) frame. *How many tens are there? How many ones are there? How many dots are there altogether?* Remove the ten (6 dot) frame. *How many dots are there now?*

▶ Briefly display and then screen 8 ten (10 dot) frames and a ten (3 dot) frame. *How many tens are there? How many ones are there? How many dots are there altogether?* Remove the ten (3 dot) frame. *How many dots are there now?* Check to see if you are correct.

▶ Continue with similar examples.

Purpose, Teaching and Children's Responses

▶ The activities in this teaching procedure are labelled 'subtracting to a decade' because the difference (that is, the answer dot) is a decade number.

8.4.3: Adding to a Decade: 1–5

▶ Place out 5 ten (10 dot) frames and a ten (8 dot) frame. *How many dots are there altogether? How many more to make 60?*

▶ Briefly display and then screen 2 ten (10 dot) frames and a ten (9 dot) frame. *How many dots are there altogether? How many more dots to make 30?*

▶ Briefly display and then screen 4 ten (10 dot) frames and a ten (5 dot) frame. *How many dots are there altogether? How many more dots to make 50?*

▶ Continue with similar examples.

Purpose, Teaching and Children's Responses

▶ These activities involve adding a number in the range 1 to 5 to obtain a decade number.

▶ Adding a number in the range 1 to 5 to obtain a decade number is likely to be easier than adding a number in the range 6 to 9 to obtain a decade number.

▶ Encourage children to use strategies other than counting by ones, for example, strategies that involve reasoning with visualized patterns.

8.4.4: Subtracting from a Decade: 1–5

▶ Place out 4 ten (10 dot) frames. *How many dots are there?* Screen 2 dots on one of the ten (10 dot) frames. *I have screened 2 dots. How many dots are there now?*

▶ Briefly display and then screen 3 ten (10 dot) frames. *How many dots are there?* Ask the child to look away while placing a small screen over 3 dots. *I have taken away 3 dots. How many dots are there now?* Check to see if you are correct.

▶ Briefly display and then screen 9 ten (10 dot) frames. *How many dots are there?* Ask the child to look away while placing a small screen over 4 dots. *I have taken away 4 dots. How many dots are there now?* Check to see if you are correct.

▶ Continue with similar examples.

Purpose, Teaching and Children's Responses

▶ These activities involve subtracting a number in the range 1 to 5 from a decade number.
▶ Subtracting a number in the range 1 to 5 from a decade number is likely to be easier than subtracting a number in the range 6 to 9 from a decade number.
▶ As above, encourage children to use strategies other than counting by ones.

8.4.5: Adding to a Decade: 6–9

▶ Place out 5 ten (10 dot) frames and a ten (2 dot) frame. *How many dots are there altogether? How many more to make 60?*
▶ Briefly display and then screen 2 ten (10 dot) frames and a ten (4 dot) frame. *How many dots are there altogether? How many more dots to make 30?*
▶ Briefly display and then screen 4 ten (10 dot) frames and a ten (1 dot) frame. *How many dots are there altogether? How many more dots to make 50?*
▶ Continue with similar examples.

Purpose, Teaching and Children's Responses

▶ As above, encourage children to use strategies other than counting by ones.
▶ Children might attempt to reason from a known result, for example using 2 and 8 make 10 to determine that 52 and 8 make 60.

8.4.6: Subtracting from a Decade: 6–9

▶ Place out 4 ten (10 dot) frames. *How many dots are there?* Screen 6 dots on one of the ten (10 dot) frames. *I have screened 6 dots. How many dots are there now?*
▶ Briefly display and then screen 8 ten (10 dot) frames. *How many dots are there?* Ask the child to look away while placing a small screen over 7 dots. *I have taken away 7 dots. How many dots are there now?* Check to see if you are correct.
▶ Briefly display and then screen 6 ten (10 dot) frames. *How many dots are there?* Ask the child to look away while placing a small screen over 8 dots. *I have taken away 8 dots. How many dots are there now?* Check to see if you are correct.
▶ Continue with similar examples.

Purpose, Teaching and Children's Responses

▶ As above, encourage children to use strategies other than counting by ones.
▶ Children might attempt to reason from a known result, for example using 10 take away 6 make 4 to determine that 40 take away 6 make 34.

MATERIALS

10× ten frames with 10 dots – called ten (10 dot) frame.

Ten frames with 1, 2, … 9 dots – called ten (1 dot) frame and so on.
Rectangles of cardboard for screening bundles and sticks.
Small rectangles for screening parts of a ten frame.

Partitioning

Key Topic 8.5: Addition and Subtraction to 20, Using 5 and 10

Purpose: To develop facility with addition and subtraction in the range 1 to 20, using grouping by 5 and 10.

LINKS TO LFIN

Main links: SEAL, Stage 5.
Other links: Tens and Ones, Level 2.

TEACHING PROCEDURES

These procedures involve using the arithmetic rack. Each procedure can be carried out in the following phases:

▶ The child uses the arithmetic rack.
▶ The teacher poses tasks using the overhead projector (OHP) arithmetic rack (or the regular rack with screening) and the child determines answers by looking at the rack.
▶ The teacher poses tasks by flashing the OHP arithmetic rack (or the regular rack with screening) and the child determines answers without looking at the rack, and then looks at the rack to confirm answers.

8.5.1: Building Numbers 6–10

▶ *Move 5 beads on the upper row. Move 1 more on the upper row. How many altogether?*
▶ *Move 5 beads on the upper row. Move 2 more on the upper row. How many altogether?*
▶ Similarly for 5 and 3, 5 and 4, 5 and 5.
▶ Repeat for 5 and 1, 5 and 2, and so on in random order.

Purpose, Teaching and Children's Responses

▶ An important goal is for children to develop strategies that do not involve counting by ones.
▶ Children should come to know that, for example, 8 can be alternatively regarded as 5 and 3.

8.5.2: Doubles – 3 + 3, 4 + 4, 5 + 5

▶ *Make a group of 3 on the upper row. Make a group of 3 on the lower row.*
▶ *Move 3 over on the upper row, and 3 back on the lower. How many altogether?*
▶ Similarly for 4 + 4 and 5 + 5.
▶ Repeat in random order.

Purpose, Teaching and Children's Responses

▶ It is important to provide sufficient time for children to think and to reflect on their thinking.
▶ Children should come to know that, for example, 3 + 3 can be reorganized into 5 + 1, by moving the beads as described above.

8.5.3: Building Numbers 11–20

▶ *Move 10 beads on the upper row. Move 1 bead on the lower row. How many altogether?*
▶ *Move 10 beads on the upper row. Move 2 beads on the lower row. How many altogether?*
▶ Similarly for 10 and 3, 10 and 4, ... 10 and 10.
▶ Repeat in random order.

Purpose, Teaching and Children's Responses

▶ Encourage children to use strategies other than counting-by-ones.
▶ Children should come to know that, for example, 14 can alternatively be regarded as 10 and 4.

8.5.4: Doubles – 6 + 6 to 10 + 10

▶ *Make a group of 6 on the upper row. Make a group of 6 on the lower row. Can you see 5 and 5? How many is that? What else can you see? How many altogether?*
▶ *Move 4 over on the upper row, and 4 back on the lower. How many altogether?*
▶ Similarly for doubles from 7 + 7 to 10 + 10.
▶ Repeat in random order.

Purpose, Teaching and Children's Responses

▶ Children should come to know that, for example, 6 and 6 can alternatively be regarded as 6 and 4 and 2, or 10 and 2.

8.5.5: Doubles Plus or Minus 1

▶ *Make a group of 4 on the upper row. Make a group of 3 on the lower row. Can you work it out in your head using a double? Move 3 over on the upper row and 3 back on the lower. How many are 4 and 3?*
▶ *Make a group of 3 on the upper row. Make a group of 4 on the lower row. Can you work it out in your head using a double? Move 4 over on the upper row and 4 back on the lower. How many are 3 and 4?*
▶ Similarly for 2 and 1, 1 and 2; 3 and 2, 2 an 3; 5 and 4, 4 and 5.
▶ *Make a group of 8 on the upper row. Make a group of 7 on the lower row. Can you work it out in your head using a double? Move 2 over on the upper row and 2 back on the lower. How many are 8 and 7?*
▶ *Make a group of 7 on the upper row. Make a group of 8 on the lower row. Can you work it out in your head using a double? Move 3 over on the upper row and 3 back on the lower. How many are 7 and 8?*
▶ Similarly for 6 and 5, 5 and 6; 7 and 6, 6 an 7; 9 and 8, 8 and 9; 10 and 9, 9 and 10.
▶ Repeat in random order.

Purpose, Teaching and Children's Responses

▶ Children should come to know that, for example, 8 and 7 can alternatively be regarded as 8 and 2 and 5, or 10 and 5, or 15.

8.5.6: Addition by Going through Ten

▶ *Move over 9 on the upper row and 2 on the lower. Move over 1 on the upper and move back 1 on the lower. How many are 9 and 2?*
▶ Similarly for 9 and 3, 9 and 4, ... 9 and 9.
▶ *Move over 8 on the upper row and 3 on the lower. Move over 2 on the upper and move back 2 on the lower. How many are 8 and 3?*
▶ Similarly for 8 and 4, 8 and 5, ... 8 and 9; and 7 and 4, 7 and 5, ... 7 and 9.

Purpose, Teaching and Children's Responses

▶ Children should come to know that, for example, 8 and 4 can alternatively be regarded as 8 and 2 and 2, or 10 and 2, or 12.

8.5.7: Commutativity of Addition

▶ *Move over 2 on the upper row and 5 on the lower row. Read the numbers on each row. Move 3 over on the upper row and 3 back on the lower. Now read the numbers on each row. What do you notice?*
▶ Similarly for 4 and 9, 3 and 7, and so on.

Purpose, Teaching and Children's Responses

▶ Children should come to know that, for example, 4 and 9 is the same as 9 and 4.

8.5.8: Addition by Compensation

▶ *Move over 7 on the upper row and 9 on the lower row. Read the numbers on each row. Move 1 over on the upper row and 1 back on the lower. Now read the numbers on each row. What is 7 and 9?*
▶ Similarly for 6 and 8, 4 and 6, 9 and 7, 8 and 6, and so on.

Purpose, Teaching and Children's Responses

▶ Children should come to know that, for example, 7 and 9 is the same as 8 and 8.

8.5.9: Subtraction by Going through Ten

▶ *Move over 10 on the upper row and 3 on the lower row. What number is that? Take 4 away from 13. Take 3 on the lower row and 1 on the upper row. How many are left?*
▶ *Move over 10 on the upper row and 5 on the lower row. What number is that? Take 7 away from 15. Take 5 on the lower row and 2 on the upper row. How many are left?*
▶ Similarly for 12 take away 3, 12 take away 4, 13 take away 5, and so on.

Purpose, Teaching and Children's Responses

▶ Children should come to know that, for example, 13 take away 4 is the same as 13 take away 3 and take away 1, or 10 take away 1. Similarly, 15 take away 8 is 15 take away 5 take away 3, or 10 take away 3.

VOCABULARY

Upper row, lower row.

MATERIALS

Arithmetic racks – one per child.
Overhead projector (OHP) arithmetic rack.

ACKNOWLEDGMENT

The arithmetic rack was developed by researchers from the Freudenthal Institute (Utrecht, The Netherlands). These activities were adapted from the work of Paul Cobb and colleagues.

Key Topic 8.6: Developing Multiplication and Division

Purpose: To further develop early multiplicative and divisional strategies.

LINKS TO LFIN

Main links: Part D, Early Multiplication and Division
Other links: FNWS and BNWS, Levels 1–4; SEAL Stages 1–5.

TEACHING PROCEDURES

8.6. 1: Determining the Number in Partially Screened Equal Groups

▶   Place out three plates each containing 3 cakes. *Here are 3 plates of cakes. Now look away.* Place out two more plates of cakes under a screen. *There are 2 more plates of 3 cakes under here. How many cakes are there altogether?*

▶   Similarly 4 lots of 2 and 2 lots of 2, 2 lots of 5 and 1 lot of 5, 2 lots of 2 and 2 lots of 2.

Purpose, Teaching and Children's Responses

▶   On tasks where some groups are not screened, children will tend to use the convenient and easy strategy of counting the items by ones, starting from one, or from the number in the first group.

▶   Encourage children to use strategies other than counting-by-ones.

▶   On the first task for example, children might count by twos from 2 to 6, and then onward to 10.

8.6.2: Determining the Number in Screened Equal Groups

▶   Briefly display and then screen six 2-dot cards. *There are six lots of 2 under here. How many are there altogether?*

▶   Similarly four 5-dot cards, five 3-dot cards, and so on.

Purpose, Teaching and Children's Responses

▶   Encourage children to use strategies other than counting-by-ones.

▶   On the first task for example, children might count by twos from 2 to 12.

8.6.3: Determining the Number of Groups

▶ Similarly nine dots using 3-dot cards, 16 dots using 2-dot cards, and so on.
▶ Place three 4-dot cards under a screen. *I am using the 4-dot cards and there are 12 dots altogether. How many cards are there?*

Purpose, Teaching and Children's Responses

▶ On the first task for example, children might count by ones from one to 12, and simultaneously keep track of each set of four counts (that is, 1, 2, 3, 4; 5, 6, 7, 8; 9, 10, 11, 12 – 3). Alternatively, children might count by fours from 4 to 12 and keep track of the number of counts (that is, 4, 8, 12 – 3).

8.6.4: Determining the Number in Each Group

▶ Place seven 2-dot cards under a screen. *There are 7 cards and 14 dots altogether. What is the number on each card?*
▶ Similarly 3 cards and 15 dots, 4 cards and 8 dots, and so on.

Purpose, Teaching and Children's Responses

▶ Children might use a guess and check strategy. On the first task for example, children might guess 3 and count to see if 7 lots of 3 make 14.

8.6.5: Determining the Number in a Screened Array

▶ Briefly display and then screen a 7 × 3 array. Unscreen one row. *There are 7 rows altogether. How many dots altogether?*
▶ Similarly using the following arrays: 4 × 2, 4 × 3, 6 × 5, and so on.

Purpose, Teaching and Children's Responses

▶ On the first task for example, children might count by threes from 3 to 21.

8.6.6: Determining the Number of Rows

▶ Briefly display and then screen a 4 × 5 array. Unscreen one row. *There are 20 dots altogether. How many rows are there?*
▶ Similarly using the following arrays: 6 × 2, 8 × 3, 3 × 4, and so on.

Purpose, Teaching, and Children's Responses

▶ On the first task for example, children might count by 5s from 5 to 20, and keep track of the number of counts (that is, 5, 10, 15, 20 – 4).

8.6.7: Determining the Number in Each Row

▶ Briefly display and then screen a 3 × 4 array. *There are 12 dots altogether and there are 3 rows. How many dots in each row?*

▶ Similarly using the following arrays: 2 × 8, 6 × 3, 4 × 5, and so on.

Purpose, Teaching and Children's Responses

▶ Children might use a guess and check strategy. On the first task for example, children might guess 5 and count to see if 3 lots of 5 make 12.

VOCABULARY

Lots of. See also Key Topic 7.6.

MATERIALS

Baskets, plates, and so on for equal shares.
Small plastic objects (for example eggs, cookies) for sharing.
Dot cards for numbers 2–5.
Arrays (for example, 7 × 3, 4 × 2, 4 × 3, 6 × 5, 6 × 2, 8 × 3, 3 × 4, 2 × 8, 6 × 3, 4 × 5).

EXAMPLES OF WHOLE-CLASS LESSONS DESIGNED FOR THE CHILD AT THE COUNTING-ON STAGE

Lesson 8(a)   **Bus flashes: ten plus addition combinations to 20**
Lesson 8(b)   **Getting on and off the bus**
Lesson 8(c)   **Bus setting to 100**
Lesson 8(d)   **The blank hundred square**
Lesson 8(e)   **Arrays: promoting progress in multiplication and division strategies**
Lesson 8(f)   **Addition and subtraction pairs to 20**

LESSON 8(A)

Title:	**Bus flashes**
Purpose:	To assist children to learn the ten plus addition combinations to 20.
Links to Key Topics:	8.5
Materials:	Bus transparency for overhead projector; Ten-Plus Fish cards.
Introductory Activities:	▶ Using the overhead projector, display a blank bus.
	How many seats are there on the bus? How many seats upstairs? How many seats downstairs?

Main Focus:	▷ Bus flashes.
	Teacher flashes bus with combinations between 10 and 20 (bottom row full).
	Children respond by:
	– saying or writing down the number of passengers;
	– saying or writing down the number of unoccupied seats.
Group/Individual Tasks:	▷ Bus Fish (2 or 3 players).
	The traditional game of Fish is played using a combination of bus and numeral cards. Players make pairs of cards.
Conclusion/Summary:	▷ Using the bus transparency on the overhead projector, flash some more ten-plus combinations using the bus with the bottom row full.

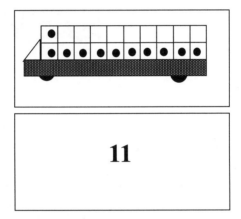

▷ Flash some ten-plus combinations using ten frames – one ten frame full and the second frame partially full.

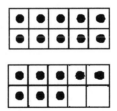

▷ Give ten-plus verbal tasks (without tens material), for example, *How many is ten and seven?*

Note: During the group activity and in the summary session, the teacher should identify those children who require further time to learn the ten-plus combinations. It would be beneficial for these children to borrow a set of Ten-Plus Fish cards to use at home.

LESSON 8(B)

Title:

Getting on and off the bus

Purpose:

The double-decker bus is used to establish addition and subtraction to 20 by adding and subtracting to 10.

Links to Key Topics:

8.4, 8.5

Material:

Bus transparency for overhead projector; bus cards and numerals cards for the Getting On and Off game and a log sheet (if required).

Introductory Activities:

▶ Bus flashes – five plus and ten plus.

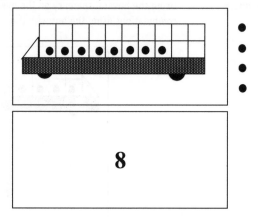

▶ Display the double-decker bus on the overhead projector.
  – Place five counters on the bottom row of the bus. *Five people get on the bus. The bus driver wants the passengers to fill the bottom row first from the front.*
  – *If three more passengers get on, how many will be on the bus?* Place three more counters on the bus then cover the bus (or it goes behind a hill or building). *How many more passengers will fit on the bottom row?*
  – While the bus is still hidden, display another four counters. *Here are another four passengers. How many will be on the top deck? How many passengers will be on the bus? How many more passengers will fit on the bus?*
  – *If two more passengers get on, how many will now be on the bus?*
  – *There are now fourteen on the bus.* Hide the bus behind an object. *If five passengers get off, how many will still be on the bus? Can you find quick ways to work this out?*

  (Have passengers clear the top deck first to emphasize subtracting to 10.)

▶ Give other tasks such as those outlined above.

Group/Individual Tasks:

Getting On and Off the Bus cards.

There are two stacks of cards. The red pack has cards showing the double-decker bus with passengers on one side and the number of passengers on the reverse.

The blue stack has numeral cards which represent passengers getting on or off the bus.

The Getting On Game

Red cards show passengers ranging from 8 to 15.

Blue cards have numerals ranging from 1 to 5. The cards can be adjusted to suit the needs of the children.

Red Passenger Cards:

Front

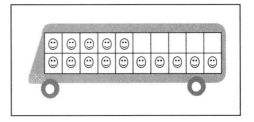

Back

$$15$$

Getting On the Bus Log Sheet		
Name: _____ Date: _____		
Passengers On The Bus	**Passengers Getting on**	**Passengers Now On The Bus**

Children play in pairs. One child selects a red bus card. The number of passengers is written on the log sheet. This same child selects a blue card and writes this number on the log sheet and then works out the number of passengers on the bus. Then the other child has a turn.

▶ The Getting Off the Bus Game

The red passenger cards range from 20 to 11 and the blue numeral cards representing people getting off the bus range from 1 to 6. The log sheet has a 'Getting Off' column in place of the 'Getting On' column.

The game is played the same as above.

Variations:

▶ As children become more proficient, the red passenger cards can be turned over, displaying the number of passengers on board rather than a picture of the bus.

▶ The number of passengers getting on or off can be generated by a die.

Conclusion/Summary: Without displaying the bus and the passengers, the whole class is asked questions such as:

There are eight people on the bus and another three get on.

- *How many more will fit on the lower deck?*

- *How many will be on the top deck?*

- *How many passengers are on the bus?*

LESSON 8(C)

Title: **Bus setting to 100**

Purpose: The tens frame (tens bus) is used to build tens and ones knowledge to 100 and adding to the next ten.

Links to Key Topics: 8.4

Materials: Bus transparency for overhead projector; bus cards; bus activity log sheet.

Introductory Activities: Counting forward by tens from 10.

Counting back from 80 by tens.

Counting forward by tens from 28. Counting back from 56 by tens.

Main Focus: − Display a tens bus.
How many people can fit on this bus?
Flash some combinations of people on a tens bus. *How many people are on the bus? How many spare seats are there?*

– Display two full tens buses.
How many people are there on these buses?
Flash combinations of people on the two buses (one full and one partly filled).
How many people are on the bus?
How many spare seats are there?

– How many tens buses?
Each child (or pair) has 10 full tens buses and a set of buses containing 1 to 9 passengers.

How many full buses are needed for 45 passengers?

Children make this using their tens buses.

– Repeat these tasks without children using buses.

▶ In these tasks we fill up a bus before we get another one!
Display then screen three full tens buses and a bus with 7 passengers (37 passengers altogether).
Place another 8 passengers next to the screened 37.
Here are another 8 passengers.
How many are needed to make another full bus?
How many passengers will there be altogether?

Display then screen 48 passengers.
Here are 48 passengers.
Place another 15 passengers next to the screen.
Here are another 15 passengers.
How many full buses will be needed?
How many passengers will there be in the partly filled bus?
How many passengers altogether?

Group/Individual Tasks: ▶ Tens Bus game.

Two stacks of cards are used. The green card stack has buses showing 31 to 71 passengers.

The red card stack has buses showing 10 to 39 passengers.

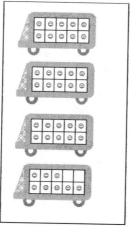

Front

Back

Both sets of cards have the numeral indicating the number of passengers on the reverse side.

Each child has a passenger log sheet.

One player selects a green card and writes the number of passengers on his log sheet. This card is turned so the number of passengers is displayed.

The same player selects a red card and writes this number on the passenger log.

The player then works out how many passengers there are altogether and records this on the log sheet.

When the log sheet is full the players can find out how many passengers there were altogether.

Bus Log Sheet Name: _____ _____			Date:	
Green card	Red card	Full Buses	Part-Filled Bus	Total Number of Passengers

▶ How many More To The Next Full Bus?

The green pack of cards mentioned above can be used, with numerals from 31 to 71 on one side and the number represented by tens buses on the other side.

The cards are placed in a stack in between the players with the numerals showing.

Children select the top card and work out how many more are needed to get to the next full bus.

To check, children look at the buses on the reverse side.

If required, the numbers and the children's answers can be logged on a record sheet.

Other Small Group Tasks:

▶ Small Group Task: Bus Fish

The traditional game of Fish is played using a combination of bus and numerals cards. Players make pairs of cards.

Conclusion/Summary:

▶ Display two bus cards and ask how many people are on the buses.

▶ Ask how many full buses are needed to carry 32 passengers. How many passengers would be on the partially filled bus?

▶ Give verbal addition tasks (that is, without using the bus cards). For example: there are 38 passengers on the buses and another 24 need to be transported.
  - *How many full buses are used for the 38 passengers?*
  - *How many passengers are in the partially filled bus?*
  - *How many more can fit on the partially filled bus?*
  - *What is the total number of passengers on the buses?*
  - *Explain how you worked out the total.*

LESSON 8(D)

Title:	**Blank hundred square**
Purpose:	To build numeral identification and number sequences to 100.
Links to Key Topics:	8.1
Materials:	Blank hundred square for overhead projector; blank hundred squares for children; full hundred squares transparencies for children to check their work.
IntroductoryActivities:	Counting forward and backward by tens from different starting points.

Display a hundred square on the overhead projector. Cover some of the numerals in a column. Pointing to a covered numeral: *What number is this?*

1	2	3	●	5	6	7	8	9	10
11	12	13	14	15	16	17	18	19	20
21	22	23	●	25	26	27	28	29	30
31	32	33	34	35	36	37	38	39	40
41	42	43	44	45	46	47	48	49	50
51	52	53	54	55	56	57	58	59	60
61	62	63	64	*What number is this?*			68	69	70
71	72	73	●	75	76	77	78	79	80
81	82	83	●	85	86	87	88	89	90
91	92	93	●	95	96	97	98	99	100

Main Focus:

▶ Display an empty hundred square. Teacher places a counter on a square and asks: *What number is this?* This is repeated for several numbers. Asks questions such as: *How did you find that number? Where did you start counting from?* (if a counting strategy was used).

I started counting from the beginning of the row.

I counted from the beginning of the square.

▶ An empty hundred square is displayed on the overhead projector. *Who can tell us how they would find where 27 would be?* A child is chosen to write in the number 27 in the correct location. *Who can tell us where 37 would be?* Methods for locating this number are discussed. A child is selected to write this number in the correct location. *Who could tell us where 57 would be? Where would 17 be?* This continues until all of the spaces in the 7 columns are filled.

Group/Individual Tasks: A selection of these whole-class activities can be used to promote counting by tens and familiarity with numbers 1 to 100.

▶ Each individual (or pair) has a blank hundred square and a transparent complete hundred square.

Teacher randomly selects a numeral from a set of numerals 1–100 (for example 45): *Write forty-five on your blank square.*

Children write 45 on their blank hundred squares.

The teacher selects other numerals at random and children write these on their blank hundred squares.

Overlaying a transparent hundred square can check these.

Note: In each of these tasks, the teacher observes how the children arrive at their answers. The class should discuss and compare their strategies.

▶ Tickle's Task (invented by Brian Tickle).

A hundred square is displayed on the overhead projector.

The square is screened by a cover with a hole displaying only one number.

Possible tasks are:

– Teacher points to the position directly below the visible numeral and asks: *What number is this?*

▶ The teacher asks: *What is the number to the right of the number below the number we can see?*

1	2	3	4	5	6	7	8	9	10
11	12	13	14	15	16	17	18	19	20
21	22	23	24	25	26	27	28	29	30
31	32	33	34	35	36	37	38	39	40
41	42	43	44	45	46	47	48	49	50
51	52	53	54	55	56	57	58	59	60
61	62	63	64	65	66	67	68	69	70
71	72	73	74	75	76	77	78	79	80
81	82	83	84	85	86	87	88	89	90
91	92	93	94	95	96	97	98	99	100

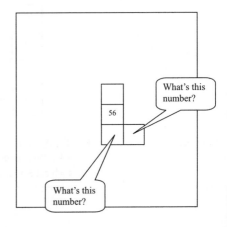

Group/Individual Tasks: A selection of these small-group activities can be used to promote counting by tens and familiarity with numbers 1 to 100.

▷ Get Ten.

Material: Each child has a blank hundred square, transparent hundred square, small numerals (1–100) to fit on blank hundred square, pencil. Children work in pairs.

A child selects a numeral from the container and places this on the blank hundred square. Alternatively, this could be written on the blank hundred square if a permanent copy of the activity is required. The other child then selects a numeral and places it on his/her blank square.

This is repeated a set number of times (say ten).

Children each check the placement of their numerals using the transparent hundred square.

Variations include:

- Children add the numbers they have in the correct locations.
- Three In A Row: The game continues until three numerals in a row have been achieved (for example, 23, 24, 25) or column (for example, 65, 75, 85).

▷ Small-group task: Find the Target Number.

On a blank hundred square, each player chooses a number and writes it in the correct location without showing the other player. This is the target number and it is circled in red.

The aim of this activity is for each player to discover the other player's target number.

Initially a player makes a guess (for example, 27). The other player states whether the target number is lower, higher or equal to the player's guess. Players keep track of these responses by writing their guesses on their blank hundred squares, and arrows are used to indicate whether the target number is higher or lower than the guess.

The game proceeds until one player or both players guess the other's target number.

▷ Partial Hundred Square.

Part of the hundred square is shown on a sheet (or on a card). Children individually or in pairs fill in the remaining squares.

Examples:

Conclusion/Summary:	▶ Display a hundred square with some of the numerals in one column screened. Pointing to each of the screened numerals, ask the class to name them.
	▶ Repeat this task using a blank hundred square.
Acknowledgment:	Tickle's Task (invented by Brian Tickle).

LESSON 8(E)

Title:	**Arrays**
Purpose:	To use arrays to promote progress in multiplication and division strategies.
Links to Key Topics:	8.6
Materials:	Array and numeral cards.
Introductory Activities:	▶ Skip counting by 2 and 5.
	▶ How many dots?
	– *There are five dots under each lid. How many dots are there altogether?*

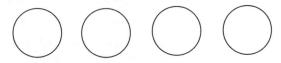

Main Focus:	▶ Display a screened array such as one of the three shown below.

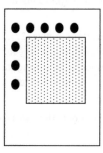 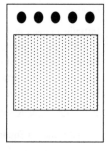 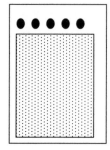

How many dots are there altogether? How did you get your answer?

There are six rows each with five dots. How many dots are there altogether? How did you get your answer?

There are 15 dots altogether. Each row has five dots. How many rows are there?

Group/Individual Tasks:	Hidden Array Fish.
	▶ Small Group
	▶ This is played the same as traditional Fish. Screened array cards are matched with numeral cards.

▶ Array and numeral cards such as those shown below for multiples of five, are used.
  − The teacher may choose to have children focus on particular multiples (for example, fives) or choose cards from a range of multiples (for example, 3, 4 and 5, and so on).

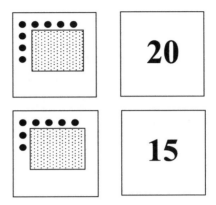

Conclusion/Summary: ▶ Whole Class.

Review by showing some more screened arrays. Give some verbal tasks:

▶ *If I have 3 rows each with 5 dots, how many are there altogether?*

▶ *Can you describe an array with 20 dots?*

LESSON 8(F)

Title:	**Addition and subtraction pairs to 20**
Purpose:	To use the double ten strip to promote addition and subtraction pairs to 20.
Links to Key Topics:	8.4
Materials:	Child double ten strips; Bingo cards; Snap cards.
Introductory Activities:	▶ Flash (that is, briefly display and then conceal) an empty ten strip.

What did you see?

▶ **Empty tens strip**

Flash (that is, briefly display and then conceal) two tens strips as shown below.

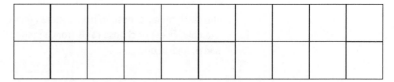

What did you see?

▶ **Double tens strip**

The fives could also be highlighted like this:

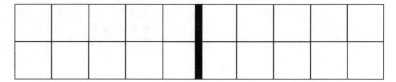

What did you see?

▶ **Double tens strip – ten plus patterns.**

Flash (that is, briefly display and then conceal) ten plus patterns:

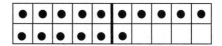

What did you see?

Main Focus: ▶ Tens frame flashes.

Teacher flashes two ten frames with ten plus combinations. Children respond in any or all of these ways:
- saying the number of dots;
- making the number of dots using finger pattern;
- making the pattern on a child double ten strip;
- writing down their response;
- showing numeral card to indicate response.

▶ Whole Class: Double Tens Strip Bingo

Children are each given Bingo cards that have numerals from 10 to 20 as shown below: The traditional game of Bingo is played with the teacher flashing ten plus combinations to 20.

15	19	15	10
18	11	20	16

Group/Individual Tasks: Matching pairs Snap.

One child has the blue stack of cards and the other child has the pink stack of cards.

The traditional game of Snap is played, with children matching double ten strip cards with numeral cards.

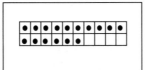

Conclusion/Summary: To the whole class give questions such as:
- What is ten and six?
- What is nine and six? How did you get your answer?
- What is 16 take away ten?
- What is 16 take away six?

9

Teaching the Facile Child

This chapter focusses on teaching children at the Facile Stage on the SEAL (Stages of Early Arithmetical Learning – see Chapter 1). First, a detailed description of the knowledge and strategies typical of this stage is provided. This is followed by explanations of the six key topics which can provide a basis for teaching the child at the Facile Stage. The six key topics consist of a total of 34 teaching procedures.

THE TYPICAL FACILE CHILD

This section provides an overview of a typical child at the Facile Stage, that is at Stage 5 of the Stages of Early Arithmetical Learning. Table 9.1 sets out the stage and indicative levels for the counting-on child on the models pertaining to FNWSs, BNWSs, Numeral Identification, Tens and Ones, and Multiplication and Division. The overview discusses the six aspects of early number knowledge listed in Table 9.1, and also grouping by fives and tens.

Table 9.1 Stage and levels of a typical facile child

Model	Stage/level
Stage of Early Arithmetical Learning (SEAL)	5
Level of Forward Number Word Sequences (FNWSs)	5
Level of Backward Number Word Sequences (BNWSs)	5
Level of Numeral Identification	4
Level of Tens and Ones Knowledge	2
Level of Early Multiplication and Division Knowledge	3

Stage of Early Arithmetical Learning

The child at the facile stage typically has developed strong facility in a range of aspects of the Learning Framework in Number. Facile arithmetical strategies might include knowing a range of doubles and using doubles to work out other facts, for example, using 5 + 5 to work out 4 + 5. These strategies can also incorporate knowledge of the numbers in the teens in terms of a ten and ones, for example, the child uses knowledge of 3 + 3 to work out 13 + 3. Understanding of the inverse relationship between addition and subtraction is also common at this stage. Thus the child might use knowledge of 9 + 3 to work out 12 – 3, or knowledge of 10 + 5 to work out 15 – 11.

Grouping by Fives and Tens

The child at this stage might use grouping by fives and tens when adding or subtracting in the range 1 to 20 (for example, 4 + 3 is found by partitioning 3 into 1 and 2, and then adding 2 to 5, and 13 – 8 is found by partitioning 8 into 3 and 5, and then subtracting 5 from 10). As well, they might solve without counting by ones, additions such as 40 + 8 and subtractions such as 28 + x = 30, 87 – x = 80, and 30 – 2. Finally, the child might use the strategy of adding to 10, for addition (for example, 18 + 6 is found by finding the complement of 8 in 10, that is, 2, and then partitioning 6 into 2 and 4, and then adding 4 to 20), and subtracting to 10 for subtraction (for example, 42 – 6 is found by partitioning 6 into 2 and 4, and then subtracting 4 from 40).

FNWSs, BNWSs and Numeral Identification

The child at this stage typically is facile with FNWSs and BNWSs in the range 1 to 100 and beyond, and can count forwards and backwards by 2s, 10s, 5s, 3s and 4s. The child can recognize and identify numerals in the range 1 to 100. In many cases the child can also recognize and identify many 3-digit numerals as well, although numerals with the digit '0' in the right-hand or middle place (for example, 620, 808) are likely to be more difficult.

Tens and Ones

The child at this stage is likely to be able to increment and decrement in the range 1 to 100, by tens on and off the decade (for example, 40 and 10, 70 less 10, 32 and 10, 88 less 10), and by tens and ones on and off the decade (for example, 32 and 2 tens and 3 ones, 87 less 2 tens less 2 ones).

Early Multiplication and Division

The child at this stage is likely to be able to use skip counting and repeated addition or subtraction to solve multiplicative and divisional tasks involving screened equal groups and screened arrays. Examples of these tasks are determining the number of items altogether in equal groups or an array; and given the number of items altogether determining the number of groups, the number in each group or the number in each row of an array.

The Way Forward

The child at this stage can extend their ability to count by tens and ones on and off the decade in the range 1 to 100 to the range from 100 to 1000, and can learn to count forwards and backwards by 100s on and off the 100, and on and off the decade. The child can also learn to convert non-canonical forms of 2-digit numbers to their canonical forms and vice versa (for example, convert 4 tens and 26 ones to 66, and convert 66 to 5 tens and 16 ones). As well, the child is ready to extend their initial knowledge involving addition and subtraction beyond 20 (for example, involving a 1-digit number and a decade number). This is extended to higher decade addition and subtraction (that is, involving a 1-digit number and a 2-digit number). These are further extended to addition and subtraction involving two 2-digit numbers without and with regrouping in counting-based settings (for example, the empty number line) and in collections-based settings (for example, bundling sticks). Finally, the child's knowledge of multiplication and division can be extended to the development of a range of strategies such as using a known fact, using multiplication to solve division, using the commutativity of multiplication, and combining and partitioning involving sets of equal groups.

The next section contains six key topics and 34 teaching procedures which form the basis of an appropriate instructional programme for the facile child.

Key Topic 9.1: Counting by 10s and 100s

Purpose: To develop facility with counting forwards and backwards by tens off the decade, and counting by 100s on and off the 100, and on and off the decade.

LINKS TO LFIN

Main links: Part A, Tens and Ones.
Other links: FNWS and BNWS, Levels 1–5.

TEACHING PROCEDURES

9.1.1: Counting by 10s off the Decade

▶   Place out three 10-strips. *How many dots are there?* Place out a 3-strip. *How many dots are there now? Now I am going to put out another ten. How many dots are there now? And another ten. How many now?* And so on.

▶   Place out nine 10-strips and 3 ones. *There are nine tens and three ones. How many dots altogether?* Remove one 10-strip. *Now how many dots are there?* Remove another 10-strip. *Now how many dots are there?* (93, 83, ... 3).

▶   Place out three 10-strips and a 3-strip. *How many dots are there?* Place a screen over the strips and place a 10-strip beside the screen. *Here is another ten, how many is that altogether?* Place out another 10-strip. *How many now?* (33, 43, ... 93).

▶   Remove the screen and put all of the strips together. Cover the nine 10-strips and the 3-strip. Remove and display one 10-strip. *There were 93 dots, and I've taken away one ten. How many dots are there now?* Remove and display another 10-strip. *How many dots are there now?* (93, 83, ...3).

Purpose, Teaching and Children's Responses

▶   The settings used in this key topic emphasize the unitary aspect rather than the composite aspect of numbers. Thus the 100-square, 10-strip and strips for numbers in the range 2 to 9, all serve to emphasize numbers greater than one as units.

▶   The technique of screening the strips has the purpose of advancing children's conceptually based strategies (that is, in contrast to perceptually based strategies where the child reasons in situations where they can see the materials). In the situations involving screening, children are likely to reason in terms of visualized images of the materials. This can serve to strengthen children's conceptually based thinking.

▶   Children are likely to have more difficulty with counting backwards than with counting forwards.

9.1.2: Counting by 100s to 1,000

▶ Place out one 100-square. *How many dots are there?* Place out another 100-square. *How many dots are there now?* (100, 200, … 1000).

▶ *Now I'm going to take one 100-square away.* Remove one 100-square. *How many dots are there now?* Remove another 100-square. *How many dots are there now?* (900, 800, … 0).

▶ Place out one 100-square and screen it. *How many dots are under the screen?* Place another 100-square under the screen. *How many dots are under the screen now?* (100, 200, … 1000).

▶ Place out ten 100-squares. *How many dots are there?* Place a screen over the ten 100-squares. *Now I'm going to take one 100-square away.* Remove one 100-square. *How many dots are there now?* Remove another 100-square. *How many dots are there now?* (900, 800, … 0).

Purpose, Teaching and Children's Responses

▶ The counting in this teaching procedure can be called 'counting by 100s on the 100' because each number counted is a number of whole 100s (that is, 100, 200, 300, and so on). The term 'counting by 100s off the 100' is used in cases where each number counted is not a whole number of 100s (for example, 120, 220, 320, and so on).

9.1.3: Counting by 10s beyond 100

▶ Place out one 100-square. *How many dots are there?* Place out one 10-strip. *How many dots are there now?* Place out another 10-strip. *How many dots are there now?* (100, 110, … 200).

▶ *Now I'm going to take one 10-strip away.* Remove one 10-strip. *How many dots are there now?* Remove another 10-strip. *How many dots are there now?* (190, 180, … 100).

▶ Place out three 100-squares. *How many dots are there?* Place out one 10-strip. *How many dots are there now?* (300, 310, … 400).

▶ *Now I'm going to take one 10-strip away.* Remove one 10-strip. *How many dots are there now?* Remove another 10-strip. *How many dots are there now?* (390, 380, … 300).

▶ Place out six 100-squares. *How many dots are there?* Place a screen over the six 100-squares. Place out one 10-strip and then place it under the screen. *How many dots are there now?* (600, 610, … 700).

▶ *Now I'm going to take one 10-strip away.* Remove and display one 10-strip. *How many dots are there now?* Remove and display another 10-strip. *How many dots are there now?* (690, 680, … 600).

Purpose, Teaching and Children's Responses

▶ When counting by 10s beyond 100 children are likely to have more difficulty with the number words at the beginning or end of each 100 (for example, 110 – after 100, 120 – after 110, after 190 – 200), because linguistic patterns are not so evident with these number words.

9.1.4: Counting by 100s off the 100

▶ Place out one 100-square and two 10-strips. *How many dots are there?* Place out another 100-square. *How many dots are there now?* (120, 220, … 920).

▶ Remove one 100-square. *How many dots are there now?* Remove another 100-square. *How many dots are there now?* (820, 720, … 20).

▶ Place out five 10-strips. *How many dots are there?* Place a screen over the five 10-strips. Place out a 100-square and then place it under the screen. *How many dots are there now?* Place out another 100-square and then place it under the screen. *How many dots are there now?* (50, 150, … 950).

▶ Place out nine 100-squares and seven 10-strips. *How many dots are there?* Place a screen over the squares and strips, and then remove one 100-square. *How many dots are there now?* Remove another 100-square. *How many dots are there now?* (970, 870, … 70).

Purpose, Teaching and Children's Responses

▶ When using the screening technique it is important to give children sufficient time to think about the number of dots remaining, and to reflect on their thinking.

▶ In the case of counting backwards, children are likely to use a strategy involving several discrete steps, for example, first figuring out how many 100-squares remain and then figuring out the number word corresponding to the number of dots in all remaining. This strategy can be contrasted with an adult-like strategy of thinking immediately of the corresponding sequence of number words.

9.1.5: Counting by 100s off the Hundred and off the Decade

▶ Place out three 10-strips and a 2-strip. *How many dots are there?* Place a screen over the strips. Place out a 100-square and then place it under the screen. *How many dots are there now?* Place out another 100-square and then place it under the screen. *How many dots are there now?* (32, 132, 232, … 932).

▶ Place out nine 100-squares, eight 10-strips and a 1-strip. *How many dots are there?* Place a screen over the squares and strips, and then remove one 100-square. *How many dots are there now?* Remove another 100-square. *How many dots are there now?* (981, 881, … 81).

Purpose, Teaching and Children's Responses

▶ Children will become aware of linguistic patterns when counting by 100s in this way, particularly in the case of counting forwards.

9.1.6: Counting by 10s beyond 100 off the Decade

▶ Place out one 100-square, two 10-strips and an 8-strip. *How many dots are there?* Place a screen over the dots and place another 10-strip under the screen. *Here is another ten, how many is that altogether?* Place another 10-strip under the screen. *How many now?* (128, 138, … 308).

▶ Place out nine 100-squares, five 10-strips and a 4-strip. *How many dots are there?* Place a screen over the dots and remove one 10-strip from under the screen. *I have taken away ten, how many are left?* Remove another 10-strip from under the screen. *How many are left now?* (954, 944, … 794).

Purpose, Teaching and Children's Responses

▶ As in 9.1.3 above, children are likely to have more difficulty with the number words at the beginning or end of each 100 because linguistic patterns are less evident there.

VOCABULARY

10-strip, 100-square.

MATERIALS

1-strip, 2-strip, … 9-strip.

10 × 10-strips (a 10-strip is a strip containing a row of equally spaced dots).

10 × 100-squares (a 100-square in this case is a square consisting of ten rows of ten dots and is thus equivalent to ten 10-strips).

Screen (rectangular piece of cardboard).

ACKNOWLEDGMENT

These activities were developed in the Mathematics Recovery project.

Key Topic 9.2: 2-Digit Addition and Subtraction through Counting

Purpose: To develop counting-based strategies for 2-digit addition and subtraction.

LINKS TO LFIN

Main links: SEAL, Stage 5; Tens and Ones, Levels 2–3.

Other links: FNWS and BNWS, Levels 1–5.

TEACHING PROCEDURES

These procedures involve use of the empty number line (ENL).

9.2.1: Adding Tens to a 2-Digit Number

▶   Draw an ENL with a mark near the left-hand end. Write 32 + 40 under the ENL. *Start from 32 and add 40. Tell me how you worked it out. I will write marks on the ENL to show how you worked it out.*

▶   Write 25 + 20. *This time, you write marks on the ENL to show how you worked it out.*

▶   Similarly 46 + 30, 12 + 80, 53 + 40, and so on.

Purpose, Teaching and Children's Responses

▶   On the first task (32 + 40), the teacher writes marks on the ENL in accordance with the child's strategy. This might involve hopping as follows: 32 to 42 to 52 to 62 to 72 or 32 to 72.

▶   In this key topic the ENL provides a setting in which children initially can demonstrate their strategies. As their thinking advances the ENL supports child thinking and the development of more advanced strategies. For example, the child might initially solve 32 + 40 by counting forward in four jumps (that is, 32, 42, 52, 62, 72). Later the child might solve 32 + 40 by making one jump (that is, 32, 72).

▶ The ENL is likely to support 'jump' strategies for adding and subtracting two 2-digit numbers. Jump strategies, in the case of addition, are those where the child does not separate the first addend into tens and ones (for example, in the case of 32 + 40 the child does not first add 30 to 40). Similarly, in the case of subtraction, jump strategies are those where the child does not separate the minuend into tens and ones.

Adding tens to a 2-digit number (1)

Adding tens to a 2-digit number (2)

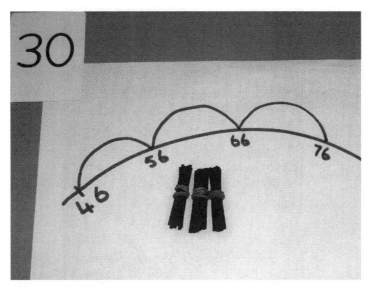

Adding tens to a 2-digit number (3)

Adding tens to a 2-digit number (4)

9.2.2: Adding Two 2-Digit Numbers Without Regrouping

▶ Draw an ENL with a mark near the left-hand end. Write 31 + 53 under the ENL. *Start from 31 and add 53. Write marks on the ENL to show how you worked this out.*

▶ *Can you work it out another way? Draw another ENL and use that to work it out another way.*

Adding two 2-digit numbers without regrouping (1)

Adding two 2-digit numbers without regrouping (2)

Adding two 2-digit numbers without regrouping (3)

▶ *Now work this out using the ENL: 34 + 43.*
▶ *Now draw another ENL and try to work it out another way.*
▶ Similarly 13 + 36, 27 + 51, 24 + 41.

Purpose, Teaching and Children's Responses

▶ Examples of likely jump strategies (for solving 31 + 53) are (a) 31, 34, 44, 54, 64, 74, 84; (b) 31, 41, 51, 61, 71, 81, 84; (c) 31, 34, 84; and (d) 31, 81, 84.

9.2.3: Adding Two 2-Digit Numbers With Regrouping

▶ Draw an ENL with a mark near the left-hand end. Write 29 + 63 under the ENL. *Start from 29 and add 63. Use the ENL to work this out.*
▶ *Can you work it out another way? Draw another ENL and use that to work it out another way.*
▶ *Now work this out using the ENL: 38 + 23.*
▶ *Now draw another ENL and try to work it out another way.*
▶ Similarly, 28 + 38, 35 + 55, 16 + 78.

Purpose, Teaching and Children's Responses

▶ Examples of likely jump strategies (for solving 29 + 63) are (a) 29, 30, 90, 92; (b) 29, 32, 42, 52, … 92; (c) 29, 89, 90, 92; and (d) 29, 32, 92.

9.2.4: Subtracting Tens from a 2-Digit Number

▶ *Use an ENL to work out 92 – 30. Use the ENL to explain how you worked it out. Can you work it out another way? Use the ENL to show another way this can be worked out.*

▶ Similarly, 82 – 20, 47 – 40, 63 – 20.

Purpose, Teaching and Children's Responses

▶ Examples of likely jump strategies (for solving 93 – 30) are (a) 93, 83, 73, 63; (b) 93, 73, 63; and (c) 93, 63.

9.2.5: Subtraction Involving Two 2-Digit Numbers Without Regrouping

▶ *Use an ENL to work out 84 – 32. Use the ENL to explain how you worked it out. Can you work it out another way? Use the ENL to show another way this can be worked out.*

▶ Similarly, 96 – 24, 59 – 18, 63 – 21.

Subtraction involving two 2-digit numbers

Purpose, Teaching and Children's Responses

▶ Examples of likely jump strategies (for solving 84 – 32) are (a) 84, 82, 72, 62, 52; (b) 84, 74, 64, 54, 52; (c) 84, 82, 52; and (d) 84, 54, 52.

9.2.6: Subtraction Involving Two 2-Digit Numbers With Regrouping

▶ Pose subtraction problems on the ENL involving 84 – 25. *Use the ENL to show how you worked this out. Can you work it out another way using the ENL?*

▶ Similarly, 72 – 29, 80 – 18, 70 – 17, 63 – 26.

Purpose, Teaching, and Children's Responses

▶ Examples of likely jump strategies (for solving 84 – 25) are (a) 84, 74, 64, 60, 59; (b) 84, 80, 79, 69, 59; (c) 84, 80, 70, 60, 59; and (d) 84, 64, 60, 59.

9.2.7: Missing Addend Tasks Involving Two 2-Digit Numbers

▶ *Here are two numbers on the ENL 44 and 66. Can you start from 44 and get to 66? How far is it from 44 to 66? Use the ENL to show how you worked it out. Can you work it out another way?*

▶ *Use the same method to work out 75 – 51. Start from 51 and go to 75. Use the ENL to show how you worked it out. Can you work it out another way?*

▶ *Now use the same method to work out 82 – 66. Start from 66 and go to 82. Find how far it is from 66 to 82? Use the ENL to show how you worked it out. Can you work it out another way?*

Purpose, Teaching, and Children's Responses

▶ Examples of likely jump strategies (for solving 82 – 66) are (a) 66, 76, 80, 82 – 16; (b) 66, 70, 80, 82 – 16; (c) 66, 86, 82 – 16; and (d) 66, 80, 82 – 16.

VOCABULARY

Empty number line.

MATERIALS

Board or paper for drawing of empty number lines

ACKNOWLEDGMENT

The arithmetic rack was developed by researchers from the Freudenthal Institute (Utrecht, The Netherlands). These activities were adapted from the work of Paul Cobb and colleagues.

Key Topic 9.3: Non-Canonical Forms of 2-Digit and 3-Digit Numbers

Purpose: To develop facility to associate non-canonical forms of 2-digit and 3-digit numbers with their canonical forms (for example, 4 tens and 27 ones is a non-canonical form of 67 and 6 tens and 7 ones is a canonical form of 67).

LINKS TO LFIN

Main links: SEAL, Stage 5, Tens and Ones, Levels 2–3.
Other links: FNWS and BNWS, Levels 1–5.

TEACHING PROCEDURES

These procedures involve using bundling sticks and base-10 blocks. Alternative materials of similar form could be used instead of the materials specified below.

9.3.1: 2-Digit Numbers in Canonical Form

▶ Place out 5 bundles of 10 and 6 ones. *How many bundles of 10 are there? How many ones are there? How many sticks are there altogether?*
▶ Similarly with 42, 25, 83, 17, and so on.
▶ Place 6 bundles of 10 and 4 ones under a screen. *Under the screen I have 64 sticks. How many tens are there? How many ones are there? Check to see if you are correct.*
▶ Similarly 37, 94, 50, 31, and so on.
▶ Place 4 bundles of 10 and 2 ones under the screen. *Under the screen I have 4 bundles of 10 and 2 ones. How many sticks are there altogether?*
▶ Similarly 2 tens and 6 ones, 5 tens and 8 ones, 3 tens and zero ones, 4 tens and 1 one, and so on.

Purpose, Teaching and Children's Responses

▶ An important goal of this procedure is that children develop facile knowledge of the canonical (that is, standard) tens and ones forms of 2-digit numbers.

9.3.2: 2-Digit Numbers in Non-Canonical Forms

▶ Place 2 bundles of 10 and 16 ones under the screen. *This time I have 36 sticks and there are two bundles of 10. How many ones are there? Check to see if you are correct.*
▶ Similarly with 5 bundles of 10 and 12 ones, 7 bundles of 10 and 14 ones, and so on.
▶ Place 3 bundles of 10 and 26 ones under the screen. *This time I have 56 sticks and there are 3 bundles of 10. How many ones are there? Check to see if you are correct.*
▶ Similarly with 4 bundles of 10 and 22 ones; 1 bundle of 10 and 26 ones, and so on.
▶ Place 6 bundles of 10 and 5 ones under the screen. *This time I have 65 sticks. If I have 5 bundles of 10 how many ones are there? What if I have 6 bundles of 10? Three bundles of 10? And so on.*
▶ Place 2 bundles of 10 and 42 sticks under the screen. *This time I have 2 bundles of 10 and 42 ones. How many sticks altogether? Check to see if you are correct.*
▶ Similarly 5 bundles of 10 and 31 ones, 3 bundles of 10 and 20 ones, 7 bundles of 10 and 18 ones, and so on.

Purpose, Teaching and Children's Responses

▶ Have children reorganize ones into bundles of 10 and vice versa in order to check their answers.

9.3.3: 3-Digit Numbers in Canonical Form

▶ Place out 2 hundreds, 4 tens and 6 ones. *How many hundreds are there? How many tens are there? How many ones are there? How many are there altogether?*
▶ Similarly with 521, 752, 138, and so on.
▶ Place 4 hundreds, 3 tens and 5 ones under a screen. *Under the screen I have 435. How many hundreds are there? How many tens are there? How many ones are there? Check to see if you are correct.*
▶ Similarly 193, 462, 350, 355, and so on.
▶ Place 2 hundreds, 5 tens and 1 one under a screen. *Under the screen I have 2 hundreds, 5 tens and 1 one. How many are there altogether?*
▶ Similarly 1 hundred, 4 tens and 6 ones; 8 hundreds, 1 ten and 8 ones; 3 hundreds, 2 tens and zero ones, and so on.

Purpose, Teaching and Children's Responses

▶ It is important to provide sufficient time for children to think and to reflect on their thinking.

9.3.4: 3-Digit Numbers in Non-Canonical Forms: Hundreds and Tens

▶ Place 2 hundreds and 13 tens under the screen. *This time I have 330 altogether, and there are two hundreds. How many tens are there? Check to see if you are correct.*
▶ Similarly with 4 hundreds and 23 tens, 6 hundreds and 21 tens, and so on.
▶ Place 2 hundreds and 31 tens under the screen. *This time I have 510 altogether and there are 2 hundreds. How many tens are there? Check to see if you are correct.*
▶ Similarly with 1 hundred and 35 tens, 6 hundreds and 25 tens, and so on.
▶ Place 7 hundreds and 4 tens under the screen. *This time I have 740 altogether. If I have 5 hundreds how many tens are there? What if I have 4 hundreds? Two hundreds?* And so on.
▶ Place 6 hundreds and 25 tens under the screen. *This time I have 6 hundreds and 25 tens. How many altogether? Check to see if you are correct?*
▶ Similarly 1 hundred and 25 tens, 4 hundreds and 15 tens, and so on.

Purpose, Teaching and Children's Responses

▶ The activities above can be extended to include ones in the range from 1 to 9 as well as hundreds and tens.
▶ Have children reorganize hundreds into tens and vice versa in order to check their answers.

9.3.5: 3-Digit Numbers in Non-Canonical Forms: Hundreds, Tens and Ones

▶ Place out 4 hundreds, 15 tens and 13 ones. *This time I have 4 hundreds, 15 tens and 13 ones. How many altogether?*
▶ Similarly with 2 hundreds, 13 tens and 12 ones, and so on.
▶ Place 4 hundreds, 5 tens and 6 ones under the screen. *This time I have 456. If there are 3 hundreds how many tens and ones could there be?*
▶ Similarly with 8 hundreds, 1 ten and 2 ones, and so on.

Purpose, Teaching and Children's Responses

▶ Have children reorganize hundreds into tens and vice versa, and tens into ones and vice versa in order to check their answers.

MATERIALS

Bundling sticks in 100s, 10s and 1s.
Bands for tying bundling sticks into groups of 10 and groups of 10 × 10 (that is, 10 lots of 10).
Base ten blocks (100s, 10s and 1s).

Key Topic 9.4: 2-Digit Addition and Subtraction through Collections
Purpose: To develop collections-based strategies for 2-digit addition and subtraction.

LINKS TO LFIN

Main links: SEAL, Stage 5, Tens and Ones, Levels 2–3.
Other links: FNWS and BNWS, Levels 1–5.

TEACHING PROCEDURES

These procedures involve use of bundling sticks in two colors.

9.4.1: Adding Two 2-Digit Numbers Using Visible Collections

▶ Using red sticks place out 4 bundles of ten and 2 ones. *How many red sticks are there?* Using green sticks place out 2 bundles of ten and 2 ones. *How many green sticks are there? How many sticks are there altogether? Explain how you worked that out.*
▶ Similarly 28 and 24. *How many red sticks are there? How many green sticks are there? How many sticks are there altogether? Explain how you worked that out. Use the sticks to show how you worked that out.*
▶ Similarly 51 + 34, 59 + 26, 77 + 24, and so on.

Purpose, Teaching and Children's Responses

▶ Observe closely to determine the child's strategy.
▶ In the case of visible collections, some children will count by ones when counting the ones and by tens when counting the tens, because it is convenient to do this. Such children might use more advanced strategies on tasks in the teaching procedures below, because the material is no longer visually available.
▶ A common strategy is to work separately with the tens and ones, and combine tens and ones at the end. This is called the 'split' strategy.
▶ Some children using the split strategy will combine the ones first and others will combine the tens first.

▶ Some children will start from one number and increment by tens or by ones. This is called the 'jump' strategy.

▶ In the case of the settings used in this teaching procedure, that is, collections, children might be more likely to use a split strategy because the tens and ones are symbolized separately, that is, a bundle for each ten and a stick for each one.

▶ The jump strategy is considered to be more difficult at least initially.

▶ The jump strategy is considered to be more flexible and adaptable to additive and subtractive tasks involving carrying (that is, regrouping) from ones to tens.

▶ Color coding of the addends (that is, red and green) serves to differentiate the two addends even after the material has been reorganized.

9.4.2: Adding Two 2-Digit Numbers Using Screened Collections

▶ Briefly display and then screen 43, using red. Briefly display and then screen 23, using green. Write the number sentence: 43 + 23. *43 and 23, how many altogether? Check to see if you were correct. Explain how you worked that out. Use the sticks if you wish.*

▶ Similarly 41 + 18, 55 + 26, 39 + 29, and so on.

Purpose, Teaching and Children's Responses

▶ Observe closely to determine if children are using the split strategy or the jump strategy.

▶ An example of the split strategy for 43 + 23 is to add 3 ones and 3 ones, and then add 4 tens and 2 tens.

▶ An example of the jump strategy is to count 43, 53, 63, 64, 65, 66.

▶ Observe closely to see if children are using counting-on by ones to combine the ones and counting-on by tens to combine the tens.

9.4.3: Missing Addend Tasks Using Visible Collections

▶ Place out 40 using red. *There are 40 red sticks. Put out some green sticks so that there are 60 altogether.*

▶ Place out 33 using red. *There are 33 red sticks. Put out some green sticks so that there are 55 altogether.* Write $33 + x = 55$.

▶ Similarly with $32 + x = 51$, $79 + x = 97$, $55 + x = 80$, $36 + x = 73$, and so on.

Purpose, Teaching and Children's Responses

▶ Some children initially have difficulty in understanding missing addend tasks.

▶ These tasks enable children to enact finding a missing second addend corresponding to a stated total.

9.4.4: Missing Addend Tasks Using Screened Collections

▶ Briefly display and then screen 40, using red. Ask the child to look away while placing 12 under the screen, using green. *There were 40 red sticks. Then I put some green sticks under the screen as well. Now there are 52 sticks altogether. How many green sticks did I put under the screen?* Write the number sentence $40 + x = 52$. *Check to see if you were correct. Explain how you worked that out. Use the sticks if you wish.*

▶ Similarly $55 + x = 75$, $41 + x = 54$, $28 + x = 40$, $59 + x = 82$, and so on.

Purpose, Teaching and Children's Responses

▶   Observe closely to see if children are using a split strategy or a jump strategy.
▶   Children using a split strategy are likely to have difficulty in cases such as 59 + x = 82.

9.4.5: Subtraction Using Visible Collections

▶   Place out 86. *There are 86 sticks.* Write 86 – 20. *If I took 20 sticks away, how many would be left? Use the sticks to see if you are correct.* Write = 66 beside 86 – 20.
▶   Place out 46. *There are 46 sticks.* Write 46 – 13. *If I took 13 sticks away, how many would be left? Use the sticks to see if you are correct.* Write = 33 beside 46 – 13.
▶   Similarly with 89 – 28, 62 – 34, 70 – 25, 96 – 37, and so on.

Purpose, Teaching and Children's Responses

▶   Observe closely to see if children are using a split strategy or a jump strategy.
▶   Children using a split strategy are likely to have difficulty in cases such as 62 – 34.

9.4.6: Subtraction Using Screened Collections

▶   Briefly display and then screen 38. Write 38 – 22. *There are 38 sticks and I am going to take away 22 sticks. How many sticks would be left?* Write the child's answer beside 38 – 22. *Use the sticks to see if you are correct.*
▶   Similarly 87 – 20, 45 – 14, 72 – 28, 94 – 37, and so on.

Purpose, Teaching and Children's Responses

▶   Most of the above notes apply similarly to this teaching procedure.

MATERIALS

Red bundling sticks and green bundling sticks.
Bands for tying bundling sticks into groups of 10.

ACKNOWLEDGMENT

These activities were developed in the Mathematics Recovery project.

Key Topic 9.5: Higher Decade Addition and Subtraction

Purpose: To develop strategies for adding numbers in the range 2 to 9, to 2-digit numbers, and subtracting numbers in the range 2 to 9, from 2-digit numbers.

LINKS TO LFIN

Main links: SEAL, Stage 5; Tens and Ones, Level 3.

TEACHING PROCEDURES

9.5.1: Using Addition Facts Within the Decade

▶   Briefly display and then screen 2 red and 2 green bundling sticks. *How many red sticks? How many green sticks? How many sticks altogether?* Write 2 + 2 = 4.
▶   Place two bundles of 10 red sticks under the screen. *How many red sticks? How many green sticks? How many sticks altogether?* Write 22 + 2 = 24.
▶   Similarly, 32 + 2, ... 92 + 2.
▶   Repeat for sets of additions such as 5 + 2, 25 + 2, ... 95 + 2.
▶   *Now answer these without using the sticks. 2 + 3, 22 + 3, ... 92 + 3. Now write all of those down in order.*
▶   Repeat for sets of additions such as 4 + 4, 24 + 4, ... 94 + 4.

Purpose, Teaching and Children's Responses

▶   An important goal is for children to use 2 + 2 = 4 to work out 22 + 2, and so on. Thus children should come to see that 2 + 2, 22 + 2, 32 + 2, and so on have in common the adding of 2 ones and 2 ones.
▶   Children are more likely to see the link just described in the case of 2 + 2 and 22 + 2, compared with 2 + 2 and 12 + 2 because two in the sense of two ones is not apparent in the number name 'twelve'. Similarly, additions involving other addends in the teens (11, 13, 14, and so on) are more difficult for children to link to other additions with corresponding numbers in the ones place (for example, 3 + 5 is linked more easily to 23 + 5 than to 13 + 5).
▶   Children at this stage should have a range of strategies other than counting-by-ones. Teachers should encourage children to use strategies other than counting-by-ones to work out sums such as 2 + 2, 3 + 3, 4 + 4, and so on.

9.5.2: Using Subtraction Facts within the Decade

▶   Briefly display and then screen 5 bundling sticks. *How many sticks?* Remove and screen 2 sticks. *How many sticks left?* Write 5 – 2 = 3.
▶   Place two bundles of 10 and 5 single sticks under the screen. *How many sticks?* Remove and screen two sticks. *How many sticks left?* Write 25 – 2 = 23.
▶   Similarly, 35 – 2, ... 95 – 2.
▶   Repeat for sets of subtractions such as 9 – 6, 29 – 6, ... 99 – 6.
▶   *Now answer these without using the sticks. 7 – 4, 27 – 4, ... 97 – 4. Now write all of those down in order.*
▶   Repeat for sets of subtractions such as 6 – 3, 26 – 3, ... 96 – 3.

Purpose, Teaching and Children's Responses

▶   Encourage children to use strategies other than counting-by-ones.
▶   As above, it is important for children to use single digit subtractions to work out higher decade subtractions, that is, to see links between subtractions such as 9 – 6 and 49 – 6.

9.5.3: Using Addition Facts Across the Decade

▶ Briefly display and then screen a red ten (9) frame and a green ten (3) frame. *How many red dots? How many green dots? How many dots altogether? 9 and 3 make 12.* Write 9 + 3 = 12.

▶ *I'm going to put another 10 under here.* Place a red ten (10) frame under the screen. *Now how many red dots? How many green dots? How many dots altogether?* Write 19 + 3 = 22.

▶ Similarly, 29 + 3, ... 89 + 3.

▶ Repeat for sets of additions such as 6 + 7, 16 + 7, ... 86 + 7.

▶ *Now answer these without using the frames. 8 + 6, 18 + 6, ... 88 + 6. Now write all of those down in order.*

▶ Repeat for sets of additions such as 9 + 5, 19 + 5, ... 89 + 5.

Purpose, Teaching and Children's Responses

▶ As above, encourage children to use the initial addition to work out subsequent additions and to use strategies other than counting-by-ones.

9.5.4: Using Subtraction Facts across the Decade

▶ Place a ten (2) frame end-to-end with a ten (10) frame. Briefly display and then screen the two frames. *How many dots? I am going to cover up 4 dots. How many dots are not covered? 12 take away 4 leaves 8.* Write 12 – 4 = 8.

▶ *I'm going to put another 10 under here.* Place a ten (10) frame under the screen. *Now how many dots? I am going to cover up 4 dots? How many dots are not covered? 22 take away 4 leaves 18.* Write 22 – 4 = 18.

▶ Similarly, 32 – 4, ... 92 – 4.

▶ Repeat for sets of subtractions such as 12 – 7, 22 – 7, ... 92 – 7.

▶ *Now answer these without using the frames. 11 – 8, 21 – 8, ... 91 – 8. Now write all of those down in order.*

▶ Repeat for sets of additions such as 13 – 9, 23 – 9, ... 93 – 9.

Purpose, Teaching and Children's Responses

▶ Most of the above notes apply similarly to this teaching procedure.

MATERIALS

Red and green bundling sticks.
Rectangular pieces of cardboard for screening.
10 × ten frames with 10 dots – called ten (10) frame.
Ten frames with 1, 2, ... 9 dots – called ten (1) frame and so on.
Small rectangles for screening parts of a ten frame.

ACKNOWLEDGMENT

These activities were developed in the Mathematics Recovery project.

Key Topic 9.6: Advanced Multiplication and Division

Purpose: To develop advanced multiplicative and divisional strategies.

LINKS TO LFIN

Main links: Part D, Early Multiplication and Division.
Other links: FNWS and BNWS, Levels 1–5; SEAL, Stages 3–5.

TEACHING PROCEDURES

9.6.1: Word Problems – Multiplication

▶ *Here is a problem for you to solve. There are four children and they each have three books. How many books are there altogether?*
▶ *Try to solve this problem. There are five packets and in each packet there are two pencils. How many pencils altogether?*
▶ *There are six baskets and in each basket there are five apples. How many apples are there altogether?*
▶ Similarly with 3 × 10, 4 × 4, 2 × 8, 6 × 3, and so on.

Purpose, Teaching and Children's Responses

▶ An important goal is for children to count in multiples for all or most of their count, rather than count-by-ones.
▶ Children who count in multiples are likely to regard each equal group as a unit, rather than a composite.
▶ In the case of 6 × 3 (6 lots of 3), children might count by threes to 12 or 15 and then count-on by ones.
▶ In the case of solving 6 × 3 by counting by threes, children must keep track of the number of threes and the total number of counts.
▶ In the case of solving 6 × 3 by counting-by-ones, children must keep track of the number of ones in each three, the number of threes and the total number of counts.
▶ It is common for children to use fingers to keep track; for example, when solving 6 × 3 children might use fingers on one hand to keep track of the ones in each three and fingers on the other hand to keep track of the number of threes.

9.6.2: Word Problems – Quotition Division

▶ *Here is a problem for you to solve. Fourteen pens are put into groups of two. How many groups of two are there?*
▶ *Try to solve this problem. There are 20 cherries and the children are given 5 each. How many children are there?*
▶ *Each box of chocolates has four chocolates, and there are 12 chocolates altogether. How many boxes are there?*
▶ Similarly with 18 ÷ 3, 10 ÷ 2, 30 ÷ 10, and so on.

Purpose, Teaching and Children's Responses

▶ On these tasks children are likely to use counting strategies similar to those described in 9.6.1 above.

▶ Children who count in multiples and keep track of the number of multiples are likely to regard each equal group as a unit rather than a composite.

▶ Children might use fingers, for example, to keep track of the number of multiples.

▶ Some children might use less advanced strategies such as attempting to enact making equal groups.

9.6.3: Word Problems – Partition Division

▶ *Here is a problem for you to solve. Fifteen oranges are shared equally among three people. How many oranges does each person get?*

▶ *Try to solve this problem. Eighteen children are put into two equal groups. How many children are there in each group?*

▶ *Twenty books are shared equally among four children. How many books does each child get?*

▶ Similarly with 24 ÷ 3, 14 ÷ 2, 32 ÷ 4, and so on.

Purpose, Teaching and Children's Responses

▶ Partition problems are likely to be more difficult than quotition problems because the child cannot reason in terms of equal groups of a given number.

▶ Children might use a strategy involving estimating and checking. On 15 ÷ 3 for example, as a partition word problem, children might estimate 4 and then check by counting three fours.

9.6.4: Commutativity with Arrays

▶ Briefly display and then screen a 3 × 5 array. *How many rows did you see? How many columns did you see? How many dots altogether?*

▶ *Now I am going to turn the array around.* Turn the array through 90 degrees, and again briefly display it and then screen it. *How many rows are there now? How many columns are there now? How many dots are there now?*

▶ Unscreen the array. *Five rows of three.* Turn the array through 90 degrees. *Three rows of five. Can you see that these are the same? How many are there altogether?*

▶ Similarly using the following arrays: 2 × 6, 5 × 4, 4 × 10, and so on.

Purpose, Teaching and Children's Responses

▶ These activities can be linked with the idea of the commutativity of multiplication (the commutative principle), that is, reversing the order of two factors does not change the product.

▶ It is important to give children sufficient time to think about the tasks and to reflect on their solutions.

▶ For many children, it might not be easy for them to reason from a demonstration with the array, that 3 × 5 has the same answer as 5 × 3.

9.6.5: Combining Two Sets of Equal Groups

▶ Briefly display and then screen 5 red 2-dot cards. *Here are 5 red 2-dot cards.* Briefly display and then screen 2 green 2-dot cards. *Here are 2 green 2-dot cards. How many 2-dot cards are there altogether? How many dots are there altogether?*
▶ Briefly display and then screen 4 red 5-dot cards. *Here are 4 red 5-dot cards.* Briefly display and then screen 2 green 5-dot cards. *Here are 2 green 5-dot cards. How many 5-dot cards are there altogether? How many dots are there altogether?*
▶ Similarly using the following: 4 × 3s and 5 × 3s; 5 × 4s and 2 × 4s; 6 × 2s and 5 × 2s, and so on.

Purpose, Teaching and Children's Responses

▶ These activities can be linked with the idea of the distributive principle of multiplication (that is, a × b + a × c = a × (b + c)).
▶ As above, it is important to give children time to reflect on their thinking.

9.6.6: Partitioning Arrays

▶ Place out a 5 × 6 partitioned array card (partitioned into 3 × 6 and 2 × 6 on the reverse side), with the non-partitioned side turned up. *How many rows are there? How many columns are there?*
▶ Turn the card over. *How many red rows are there? How many green rows are there? How many rows altogether?*
▶ *We can say 3 sixes and 2 sixes makes 5 sixes. Or 5 sixes is the same as 2 sixes and 3 sixes.*
▶ Place out the 7 × 4 partitioned array card (partitioned into 5 × 4 and 2 × 4). *How many rows altogether? What does the card show? 7 fours are the same as 5 fours and 2 fours.*
▶ Similarly using the following partitioned arrays: 8 × 5 (4 × 5 and 4 × 5), 7 × 2 (6 × 2 and 1 × 2), 7 × 3 (5 × 3 and 2 × 3), and so on.

Purpose, Teaching and Children's Responses

▶ These activities can also be linked with the idea of the distributive principle of multiplication.

MATERIALS

Dot cards for numbers 2–5.
Arrays (for example, 3 × 5, 2 × 6, 5 × 4, 4 × 10).
Partitioned array cards (for example, 3 × 6 and 2 × 6, 4 × 5 and 4 × 5, 6 × 2 and 1 × 2, 5 × 3 and 2 × 3).

EXAMPLES OF WHOLE-CLASS LESSONS DESIGNED FOR THE CHILD AT THE FACILE STAGE

Lesson 9(a)   Adding/subtracting tens and ones to 1,000 – counting-based and collection-based
Lesson 9(b)   Empty number line
Lesson 9(c)   Numbers to 1,000
Lesson 9(d)   Collection-based addition and subtraction
Lesson 9(e)   Multiplication facts

LESSON 9(A)

Title:	**Adding/subtracting tens and ones to 1,000 – counting-based and collection-based**
Purpose:	To assist children to develop sophisticated strategies to add and subtract three-digit numbers.
Links to Key Topics:	9.1
Materials:	

Introductory Activities:　Counting by tens from a variety of starting points.

Counting by hundreds from hundreds. *Start counting by hundreds from 200.*

Flashing whole hundreds.

How many hundreds? How many dots?

Main Focus:

▶ Counting by hundreds from different starting points.

▶ Near hundreds.

What is 234 and 100?

What is 234 and 99?

What is 234 and 90?

If I take 100 from 456, how many dots are left?

If I take 90 from 456, how many dots are left?

If I take 99 from 456, how many dots are left?

▶ Making the next hundred.

What is 580 and 50?

What is 580 and 38?

What is 675 and 45?

▶ How many more to the next hundred?

If I have 760, how many mare do I need to make up the next full hundred square? Have children show the answer on the overhead projector.

Increase the complexity of the tasks (for example, 379, 231, and so on).

Group/Individual Tasks: ▶ How Many More?

Pairs of children.

Have cards with numerals from 100 to 1,000. The cards are placed in a stack in between the players.

Children select the top card and work out how many more are needed to get to the next hundred.

If required, the numbers and the children's answers can be logged on a record sheet.

Number	How many more?

Conclusion/Summary: ▶ Discuss the group activity. Have some pairs share the results on their record sheets.

▶ Without material, ask the class to solve tasks such as: *If I have 347, how many more do I need to make 400?*

LESSON 9(B)

Title: **Empty number line**

Purpose: Assisting children to develop sophisticated addition and subtraction strategies by the use of the empty number line.

Materials: Pink cards, blue cards and log sheet.

Links to Key Topics: 9.2

Introductory Activities: ▶ On the board or overhead projector, the teacher draws the solutions to some addition and subtraction tasks (like those shown below) and asks the class what the question was and how the children solved it.

Task A

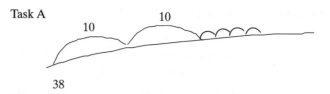

38

Task B

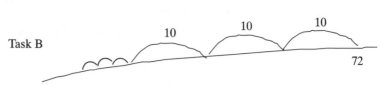

72

▶ Using ten frames (or ten strips) display and screen 48 dots. *Here are 48 dots.*

Place another 24 dots next to them. *How many dots are there altogether? Let's try to look at different ways of working it out.*

Ask children to tell how they solved this task. As solutions are given the teacher uses empty number lines to map the strategies. Some possible child strategies are shown below.

Child A: **I added ten, which made 58, then another ten which made 68 and then counted on 69, 70, 71, 72.**

Child B: **I did almost the same, 58 and then 68 but then I added two to get 70 and two more to get 72.**

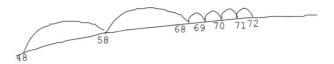

Child C: **A faster way is to start at 48 and jump 20 to 68 then two more to get 70 then two more to get 72.**

▶ Using ten frames (or ten strips) display and screen 64 dots. *Here are 64 dots.*

Remove 25 dots. *I've taken out twenty-five dots, how many are left?*

Ask children to tell how they solved this task. As solutions are given the teacher uses empty number lines to map the strategies.

Some possible child strategies are shown below.

Child A: **I came back ten to 54, then another ten to 44 then counted back 43, 42, 41, 40, 39.**

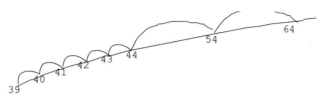

Child B: **A quicker way is to jump back by tens, 54 and 44 then come back 4 to 40 then back another one to 39.**

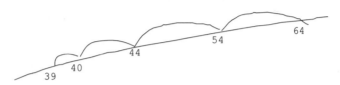

Group/Individual Tasks: Either of these activities would supplement the main focus.

▶ How many? Activity.

Children in pairs take turns to select a pink card and write the numeral displayed on the log sheet and then select a blue card and write this on the log sheet. These numbers are added using an empty number line which is drawn on the log sheet.

Pink cards have numerals from 40 to 60. Blue cards have numerals from 11 to 39.

PINK CARD	BLUE CARD	

▶ Describe the strategy.

Children in pairs or threes.

Two pink cards and two blue cards are dealt to each player. The game is played like traditional Fish with players matching a pink card showing an empty number line addition or subtraction strategy with a blue answer card as shown below.

If required, the children could write their pairs on a log sheet.

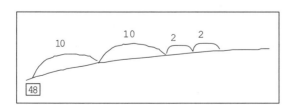

Conclusion/Summary: On the board draw some of the empty number line strategies from the 'Describe the strategy' activity and have the class discuss:

- the strategies illustrated;

- other possible strategies;

- the merits of each strategy.

LESSON 9(C)

Title: **Numbers to 1,000**

Purpose: To develop strength with numbers to 1,000.

Links to Key Topics: 9.1

Materials: Numerals for the clothes line; numeral cards for sorting.

Introductory Activities:
- ▶ *Start counting from 94.*
- ▶ *Start counting from 386.*
- ▶ *Start counting in tens from 40.*
- ▶ *Start counting in tens from 67.*
- ▶ *Start counting in hundreds from 0.*
- ▶ *Start counting in hundreds from 50.*
- ▶ *Using the hundreds square start from 4 and count by tens.*

Main Focus: Numeral Clothes Line.

A string line is strung across the room.

A set of numerals are printed on paper and folded in half so they can be placed on the line.

Assuming the task is to advance by tens, the numerals, 7, 17, 27, ... 137 are chosen.

The class is told that the smallest number is 7 and the largest number is 137.

Children are selected by the teacher to choose one of the numerals and place it on the string.

- ▶ As a variation, the end numerals can be displayed at the beginning of the task.
- ▶ This task can be repeated using different groups of numerals.

Group/Individual Tasks:
- ▶ Numeral sort.

A group of numeral cards is given to individuals or pairs of children.

The task is to arrange these cards from the smallest to the largest.

The numerals chosen should suit the learning needs of the children. The numerals could be consecutive, random, progress by tens or hundreds, and so on.

▶ Number Run Fish

A stack of cards is shuffled. Four cards are given to each player. The game is played the same as traditional Fish with players attempting to get runs of three numbers (9, 10, 11, and so on).

The cards are selected to suit the needs of those playing.

Possible groups of cards are:

Multiples of 10 or 100.

Numbers which are ten apart (for example, 4, 14, 24, ... 104).

Numbers in the hundreds (for example, 345, 346, 347, ... 400)

Conclusion/Summary:	With the whole class ask question such as:

▶ *If we are counting in tens, what is the next number after 756?*

▶ *If we are counting in tens, what is the number before 203?*

▶ *If we are counting in hundreds, what is the next number after 97?*

▶ *If we are counting in hundreds, what is the number before 567?*

LESSON 9(D)

Title:	**Collection-based addition and subtraction**
Purpose:	To develop collection-based addition and subtraction strategies.
Links to Key Topics:	9.3
Materials:	Ten frames, How Many Dots cards and log sheet.
Introductory Activities:	Each child or pair of children has ten full ten frames, a set of ten frames (1–9) and ten individual dots.

▶ *On the pink sheet of paper make 48 with your tens material. On the blue sheet of paper make 25. Write these numbers on your log sheet.*

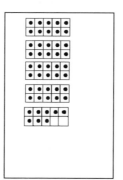 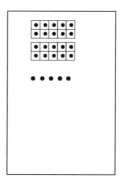

▶ *Put all the tens together on the pink sheet. How many tens are there altogether?*
Write this on your log sheet.

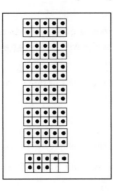 

▶ *How many dots are needed to make another ten? Move the dots to make another*
ten. On the log sheet write dawn how many tens there are now.

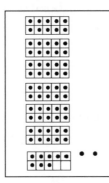 

▶ *How many dots are left? Write this on the log sheet.*

▶ *How many dots are there altogether? Write this on the log sheet.*

Main Focus:

(a) Use your ten frames to make 48 on one piece of paper.

Put 35 on the other piece of paper.

How many tens are there?

How many extra dots are there?

Can we make another ten?

How many tens are there now? How many extra dots?

How many dots altogether?

Give similar tasks.

(b) Using tens frames, display then screen 59.

There are fifty-nine dots.

Place another 23 next to the screen.

I put out another 23.

How many tens are there?

How many extra dots are there?

Can we make another ten?

How many tens are there now? How many extra dots?

How many dots altogether?

Give similar tasks.

Group/Individual Tasks: ▶   How Many Dots?

Children in pairs or individually play the How Many Dots game.

Pink cards: 36–39, 46–49, 56–59, 66–69. Blue cards: 25–27, 34–39.

Log sheet attached.

Children take turns to select a pink and a blue card, and complete the log sheet. This exercise replicates the whole-class activity.

Conclusion/Summary: ▶   With the whole class, give addition and subtraction tasks without reference to tens and ones material.

Pink Card	Blue Card	Tens	Ones	Extra Tens	Ones Left	Total

LESSON 9(E)

Title:	**Multiplication facts**
Purpose:	Using arrays to assist children in learning multiplication facts.
Links to Key Topics:	9.5
Materials:	10 × 10 multiplication square; 10 × 10 dot array.
Introductory Activities:	Display a 10 × 10 multiplication square on the overhead projector.
	Show the children how the table is read.
	Use the table to find the answer to: 5 × 6, 7 × 7, 2 × 9, and so on.
	Using a counter, cover a numeral and ask children to tell what numbers multiplied together give the covered number.
Main Focus:	▶ Display a 10 × dot array on the overhead projector.

Who could come out and show us what 5 × 8 looks like?

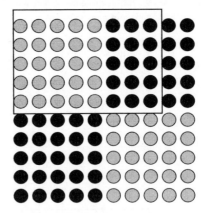

Who could come out and show us what 5 × 8 looks like?

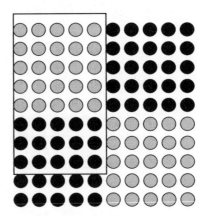

The teacher has a 5 × 8 array and turns it to demonstrate it is the same as an 8 × 5 array.

▶ *Look up 5 × 8 on your table square. Now find 8 × 5. What do you notice?*

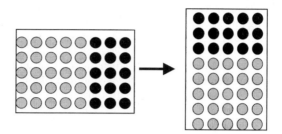

What other number pairs reverse to give the same answer when multiplied?

What is the answer to 5 × 6? What other multiplication fact does this help us with?

▶ Display another array, for example, 2 × 9. *Who can tell me what this array shows?*

If we know that 2 × 9 is 18, what is the reverse multiplication fact that we know?

Who could come out and show us why 9 × 2 has the same number of dots as 2 × 9?

▶ Give other examples of these reverse facts (the commutative property of multiplication).

Group/Individual Tasks: ▶ In pairs, children ask each other the answers to multiplication facts. The children write their names on their multiplication square sheets and exchange them with their partners. Child A has Child B's sheet and asks Child B a multiplication fact (for example, 5 ×6). If Child B answers correctly, Child A colors the 30 square. If an incorrect response is given, no square is colored.
- As children learn more multiplication facts, more squares are colored.
- Most children can achieve correct answers to multiples of 1, 2, 5 and 10.
- They can then use their knowledge of reverse facts to find facts such as 7 × 5 and 8 × 2 and so on.
- Children can focus on those facts that are still unlearnt.

Conclusion/Summary: Any of these activities would complement the main focus of this lesson:

▶ A game of Bingo using multiplication facts could be played.

▶ The teacher could ask the class a variety of multiplication facts.

▶ Children could discuss the multiplication facts with which they are confident.

▶ Some of the squares on a 10 × 10 multiplication fact square are covered and children asked what numbers multiply to give the hidden numbers.

▶ Some simple verbal tasks involving multiplication facts could be given (for example, *How many children are there if there are five teams each with three players?*).

Bibliography

Anghileri, J. (ed.) (2001) *Principles and Practices of Arithmetic Teaching*. Buckingham: Open University Press.

Aubrey, C. (1993) An investigation of the mathematical knowledge and competencies which young children bring into school, *British Educational Research Journal*, **19**(1): 37–41.

Beishuizen, M. (1993) Mental strategies and materials or models for addition and subtraction up to 100 in Dutch second grades, *Journal for Research in Mathematics Education*, **34**: 394–43.

Beishuizen, M. and Anghileri, J. (1998) Which mental strategies in the early number curriculum? A comparison of British ideas and Dutch views, *British Education Research Journal*, **34**: 519–38.

Blanck, G. (1990) Vygotsky: the man and his cause, in L.C. Moll (ed.), *Vygotsky and Education: Instructional Implications and Applications of Sociohistorical Psychology*. Cambridge: Cambridge University Press. pp. 31–58.

Bobis, J., Clarke, B., Clarke, D., Thomas, G., Wright, R., Young-Loveridge, J. and Gould, P. (2005) Supporting teachers in the development of young children's mathematical thinking: three large-scale cases, *Mathematics Education Research Journal*, **16**(3): 27–57.

Brophy, J. and Good, T.L. (1986) Teacher behaviour and student achievement, in M.C. Wittrock (ed.), *Handbook of Research on Teaching*, 3rd edn. New York: Macmillan. pp. 363–7.

Carpenter, T.R, Fennema, E., Franke, M.L., Levi, L. and Empson, S.B. (1999) *Children's Mathematics: Cognitively Guided Instruction*. Portsmouth, NH: Heinemann.

Cazden, C. (1986) Classroom discourse, in M.C. Wittrock (ed.), *Handbook of Research on Teaching*, 3rd edn. New York: Macmillan. pp. 438–53.

Cobb, P. (1991) Reconstructing elementary school mathematics, *Focus on Learning Problems in Mathematics*, **13**(3): 3–33.

Cobb, P. and Bauersfeld, H. (eds) (1995) *The Emergence of Mathematical Meaning: Interaction in Classroom Cultures*. Hillsdale, NJ: Lawrence Erlbaum.

Cobb, P. and Wheatley, G. (1988) Children's initial understandings of ten, *Focus on Learning Problems in Mathematics*, **10**(3): 1–36.

Cobb, P., Boufi, A., McClain, K. and Whitenack, J. (1997a) Reflective discourse and collective reflection, *Journal For Research in Mathematics Education*, **38**: 358–77.

Cobb, P., Gravemeijer, K., Yackel, E., McClain, K. and Whitenack, J. (1997b) Symbolizing and mathematizing: the emergence of chains of signification in one first-grade classroom, in D. Kirshner and J.A. Whitson (eds), *Situated Cognition Theory: Social, Semiotic, and Neurological Perspectives*. Mahwah, NJ: Lawrence Erlbaum. pp. 151–233.

Cobb, P., McClain, K., Whitenack, J. and Estes, B. (1995) Supporting young children's development of mathematical power, in A. Richards (ed.), *Proceedings of the Fifteenth Biennial Conference of the Australian Association of Mathematics Teachers*. Darwin: Australian Association of Mathematics Teachers. pp. 1–11.

Cobb, P., Wood, T. and Yackel, E. (1991) A constructivist approach to second grade mathematics, in E. von Glasersfeld (ed.), *Radical Constructivism in Mathematics Education*. Dordrecht: Kluwer. pp. 157–76.

Cobb, P., Wood, T. and Yackel, E. (1992) Interaction and learning in classroom situations, *Educational Studies in Mathematics*, **33**: 99–133.

Cobb, P., Yackel, E. and K. McClain, K. (eds) (2000) *Symbolizing and Communicating and in Mathematics Classrooms: Perspectives on Discourse, Tools, and Instructional Design*. Mahwah, NJ: Lawrence Erlbaum.

Department for Education and Employment (DfEE) (1999a) *Framework for Teaching Mathematics from Reception to Year 6*. Cambridge: Cambridge University Press.

Department for Education and Employment (DfEE) (1999b) *National Numeracy Strategy: Mathematical Vocabulary*. London: DfEE.

Department for Education and Skills (DfES) (2004) *What Works for Children with Mathematical Difficulties*, Research Brief RB554. London: DfES.

Department for Education and Skills (DfES) (2005) *Targeting Support: implementing interventions for children with significant difficulties in mathematics, Primary National Strategy, 1983–2005*. London: DfES.

Department of Education, Training and Youth Affairs DETYA (2000) *Mapping the Territory: Primary Students with Learning Difficulties*. Canberra: DETYA.

Dowker, A.D. (2003) Interventions in numeracy: individualized approaches, in I. Thompson (ed.), *Enhancing primary mathematics teaching*. Maidenhead: Open University Press. pp. 127–35.

Dowker, A.D. (2004) *Children with Difficulties in Mathematics: What works?* London: DfES.

Dowker, A.D. (2005) *Individual Differences in Arithmetic: Implications for Psychology, Neuroscience and Education*. Hove: Psychology Press.

Fuson, K.C., Wearne, D., Hiebert, J., Human, P., Olivier, A., Carpenter, T. and Fenema, E. (1997) Children's conceptual structure for multidigit numbers and methods of multidigit addition and subtraction, *Journal for Research in Mathematics Education*, 38: 130–63.

Gravemeijer, K.P.E. (1994) *Developing Realistic Mathematics Education*. Utrecht: CD-B Press.

Gravemeijer, K.P.E., Cobb, P., Bowers, J. and Whitenack, J. (2000) Symbolizing, modeling, and instructional design, in P. Cobb, E. Yackel and K. McClain (eds), *Symbolizing and Communicating and in Mathematics Classrooms: Perspectives on Discourse, Tools, and Instructional Design*. Mahwah, NJ: Erlbaum. pp. 335–73.

Gray, E.M. (1991) An analysis of diverging approaches to simple arithmetic: preference and its consequences, *Educational Studies in Mathematics*, 33: 551–74.

Hird, D. (2004) An evaluation of the impact of the Mathematics Recovery Programme on the staff and pupils in Sefton LEA, unpublished MEd thesis, University of Liverpool.

Kamii, C. (1985) *Young Children Reinvent Arithmetic*. New York: Teachers College Press.

Kamii, C. (1986) Place value: an explanation of its difficulty and educational implications for the primary grades, *Journal of Research in Early Childhood Education*, 1: 75–86.

Krutetski, V.A. (1976) *The Psychology of Mathematical Abilities in School Children*, trans. J. Teller. Chicago, IL: Chicago University Press.

Ma, L. (1999) *Knowing and Teaching Elementary Mathematics: Teachers' Understanding of Fundamental Mathematics in China and the United States*. Mahwah, NJ: Lawrence Erlbaum.

McClain, K. and Cobb, P. (1999) Supporting children's ways of reasoning about patterns and partitions, in J.V. Copley (ed.), *Mathematics in the Early Years*. Reston, VA: National Council of Teachers of Mathematics. pp. 113–18.

McMahon, B. (1998) A model for analysing one-to-one teaching in the Maths Recovery Programme, unpublished Honours thesis, Southern Cross University.

Mulligan, J.T. (1998) A research-based framework for assessing early multiplication and division, in C. Kanes, M. Goos and E. Warren (eds), *Proceedings of the 31st Annual Conference of the Mathematics Education Research Group of Australasia*, vol. 3. Brisbane: Griffith University. pp. 404–11.

Munn, P. (2005) The teacher as a learner, in R.J. Wright, A.K. Stafford and J.R. Martland (eds) *Teaching Number in the Classroom with 4-8 Year-Olds*. London: Paul Chapman Publishing. pp. 177–92.

National Council for Teachers of Mathematics (NCTM) *Principles and Standards for School Mathematics: Discussion Draft, October 1998*. Reston, VA: NCTM.

National Numeracy Project (1999) *Numeracy Lessons*. Reading: National Centre for Numeracy.

Perry, B. and Conroy, J. (1994) *Early Childhood and Primary Mathematics*. Sydney: Harcourt Brace.

Steffe, L.P. (1992) Learning stages in the construction of the number sequence, in J. Bideaud, C. Meljac and J. Fischer (eds), *Pathways to Number: Children's Developing Numerical Abilities*. Hillsdale, NJ: Lawrence Erlbaum. pp. 83–8.

Steffe, L.P. and Cobb, P. (with E. von Glasersfeld) (1988) *Construction of Arithmetic Meanings and Strategies*. New York: Springer-Verlag.

Steffe, L.P., von Glasersfeld, E., Richards, J. and Cobb, P. (1983) *Children's Counting Types: Philosophy, Theory, and Application*. New York: Praeger.

Stigler, J. and Hiebert, J. (1999) *The Teaching Gap*. New York: Free Press.

Streefland, L. (ed.) (1991) *Realistic Mathematics Education in Primary School*. Utrecht: CD-B Press.

Thomas, G. and Ward, J. (2001) *An Evaluation of the Count Me In Too Pilot Project*. Wellington: Ministry of Education. Available from Learning Media Customer Services, Box 3293 Wellington, New Zealand (item #10211).

Thompson, I. (1994) Young children's idiosyncratic written algorithms for addition, *Education Studies in Mathematics*, 36: 333–45.

Thompson, I. (ed.) (1997) *Teaching and Learning Early Number*. Buckingham: Open University Press.

Thompson, I. (ed.) (1999) *Issues in Teaching Numeracy in Primary Schools*. Buckingham: Open University Press.

Thompson, I. (ed.) (2003) *Enhancing Primary Mathematics Teaching*. Buckingham: Open University Press.

Treffers, A. (1991) Didactical background of a mathematics program for primary education, in L. Streefland (ed.), *Realistic Mathematics Education in Primary School*. Utrecht: CD-B Press.

Treffers, A. and Beishuizen, M. (1999) Realistic mathematics education in the Netherlands, in I. Thompson (ed.), *Issues in Teaching Numeracy in Primary Schools*. Buckingham: Open University Press.

Van de Walle, J.A. (2004) *Elementary and Middle School Mathematics: Teaching Developmentally*, 5th edn. Boston, MA: Pearson.

Van den Heuvel-Panhuizen, M. (1996) *Assessment and Realistic Mathematics Education*. Utrecht: Freudenthal Institute, Utrecht University.

Von Glasersfeld, E. (1982) Subitizing: the role of figural patterns in the development of numerical concepts, *Archives de Psychologie*, **50**: 191–318.

Von Glasersfeld, E. (1995) *Radical Constructivism: A Way of Knowing and Learning*. London: Falmer.

Vygotsky, L.S. (1978) *Mind in Society: The Development of Higher Psychological Processes*. Cambridge, MA: Harvard University Press. (Original work published 1934.)

Wheatley, G. and Bebout, H. (1990) in L.P. Steffe and T. Wood (eds), *Transforming Children's Mathematics Education: International Perspectives*. Hillsdale, NJ: Lawrence Erlbaum. pp. 107–11.

Wood, D., Bruner, J. and Ross, G. (1976) The role of tutoring in problem solving, *Journal of Child Psychology and Psychiatry and Allied Disciplines*, **17**: 89–100.

Wright, R.J. (1989) Numerical development in the kindergarten year: a teaching experiment, doctoral dissertation, University of Georgia (DAI, 50A, 1588; DA8919319).

Wright, R.J. (1991a) An application of the epistemology of radical constructivism to the study of learning, *Australian Educational Researcher*, **18** (1): 75–95.

Wright, R.J. (1991b) What number knowledge is possessed by children entering the kindergarten year of school? *Mathematics Education Research Journal*, **3**(1): 1–16.

Wright, R.J. (1994) A study of the numerical development of 5-year-olds and 6-year-olds, *Educational Studies in Mathematics*, **36**: 35–44.

Wright, R.J. (2000) Professional development in recovery education, in L.P. Steffe and P.W. Thompson (eds), *Radical Constructivism in Action: Building on the Pioneering Work of Ernst von Glasersfeld*. London: Falmer. pp. 134-51.

Wright, R.J. (2001). The arithmetical strategies of four 3rd-graders, in J. Bobis, B. Perry and M. Mitchelmore (eds), *Proceedings of the 25th Annual Conference of the Mathematics Education Research Group of Australasia*, vol. 2. Sydney: MERGA. (pp. 547–54).

Wright, R.J., Martland, J. and Stafford, A. (2006) *Early Numeracy: Assessment for Teaching and Intervention*, 2nd edn. London: Paul Chapman Publishing/Sage.

Wright, R.J., Stanger, G., Stafford, A. and Martland, J. (2006). *Teaching Mathematics in the Classroom to 4–8 Year-Olds*. London: Paul Chapman Publishing/Sage.

Yackel, E. (1995) Children's talk in inquiry mathematics classrooms, in P. Cobb and H. Bauersfeld (eds), *Emergence of Mathematical Meaning: Interaction in Classroom Cultures*. Hillsdale, NJ: Lawrence Erlbaum. pp. 131–62.

Yackel, E. and Cobb, P. (1996) Sociomathematical norms, argumentation, and autonomy in mathematics, *Journal for Research in Mathematics Education*, **27**, 458–77.

Yackel, E., Cobb, P. and Wood, T. (1991) Small group interactions as a source of learning opportunities in second grade mathematics, *Journal for Research in Mathematics Education*, **33**: 390–408.

Yackel, E., Cobb, P., Wood, T., Wheatley, G. and Merkel, G. (1990) The importance of social interaction in children's construction of mathematical knowledge, in T. Cooney (ed.), *1990 Yearbook of the National Council of Teachers of Mathematics*. Reston, VA: NCTM. pp. 12–21.

Young-Loveridge, J. (1989) The development of children's number concepts: the first year of school, *New Zealand Journal of Educational Studies*, **24**(1): 47–64.

Young-Loveridge, J.M. (1991) *The Development of Children's Number Concepts from Ages Five to Nine*, vols 1 and 3. Hamilton: University of Waikato.

Index